PRAISE FOR *TAMING THE MOLECULE OF MORE*

"From neurobiology to Aristotelian ethics, this book is both a practical guide for living and a philosophical meditation on the meaning of life. Entertaining, informative, and persuasive, *Taming the Molecule of More* is a must-read for anyone who wants to feel better while also doing good."
—Anna Lembke, MD, author of *Dopamine Nation: Finding Balance in the Age of Indulgence*

"Mike Long has written a book and you should read it. Not just because he's written a book—anybody can do that—but because he's written a good book, one that lays out its subject in easy-to-understand sentences for a layperson like myself and in simple yet profound thoughts. Nonfiction is not even my thing but I like Mike's book a lot; he even quotes George Bernard Shaw and that is more than enough for me. *Taming the Molecule of More* kept me interested until the very last sentence and in this world of sound bites and TikToks, that is no small achievement. Mike Long is the real deal."
—Neil LaBute, director, screenwriter, and three-time Tony Award-nominated playwright

"I absolutely loved this book! In an age where constant stimulation and immediate gratification are easily accessible, managing dopamine is crucial for maintaining our mental health, productivity, and quality of life. This book is full of so much wisdom, I found myself wanting to highlight every other line."
—Lamar Dawson, "The Dirrty King of Pop" and SiriusXM TikTok radio host

"This book elegantly provides the best scientific evidence to our current understanding of the dopaminergic reward system and describes self-care practices for the brain to be an organ at the service of the person and humanity."
—Edward Marshall, MD, PhD, Viktor E. Frankl Professor of Psychotherapy and Spirituality at the Graduate Theological Foundation

"This book spans the evolution from the dopamine molecule itself to the encompassing meaning in our life—and does it practically!"
—Professor Alfried Längle, MD, PhD, founder of Existential AnalysisColleague and biographer of Viktor Frankl

"In this perfect companion to and expansion of *The Molecule of More*, *Taming* goes further to provide actionable, usable tools and insights to control and harness clever and often misleading dopamine. I personally battle with controlling dopamine on a daily basis, but now do it with fortified armor, enhanced weaponry, and superior strategy having read this book."
—Jamie Lissow, comedian, actor, and writer

"Mike Long's sequel is an alchemy of science, wisdom, and heart. It unlocks the deep complexity of the human brain and trappings of modern life. Here's how: This book will reward you with an understanding of the role of dopamine and our often-fraught pursuit of purpose. Mike makes transformational concepts accessible through chapters that are beautifully written—and pragmatically inspiring."
—Dr. Kelly J. Otter, dean of the School of Continuing Studies at Georgetown University

"Mike has found a way to make all the great ideas about improving yourself into an actual, workable plan. I highly recommend his hands-on-and-get-on-with-it philosophy. You'll have a super freaking awesome time and stuff."
—Rodney Douglas Norman, comedian

TAMING
THE MOLECULE
OF MORE

ALSO BY MICHAEL E. LONG

The Molecule of More
(by Daniel Z. Lieberman, MD and Michael E. Long)

A Bushel of Beans and A Peck of Tomatoes
(by James Gregory with Michael E. Long)

TAMING THE MOLECULE OF MORE

A Step-by-Step Guide
to Make Dopamine
Work for You

MICHAEL E. LONG

BenBella Books, Inc.
Dallas, TX

This book is for informational purposes only. It is not intended to serve as a substitute for professional medical advice. The author and publisher specifically disclaim any and all liability arising directly or indirectly from the use of any information contained in this book. A health care professional should be consulted regarding your specific medical situation. Any product mentioned in this book does not imply endorsement of that product by the author or publisher.

Taming the Molecule of More copyright © 2025 by Michael E. Long

All rights reserved. Except in the case of brief quotations embodied in critical articles or reviews, no part of this book may be used or reproduced, stored, transmitted, or used in any manner whatsoever, including for training artificial intelligence (AI) technologies or for automated text and data mining, without prior written permission from the publisher.

BenBella Books, Inc.
8080 N. Central Expressway
Suite 1700
Dallas, TX 75206
benbellabooks.com
Send feedback to feedback@benbellabooks.com

BenBella is a federally registered trademark.

Printed in the United States of America
10 9 8 7 6 5 4 3 2 1

Library of Congress Control Number: 2024043091
ISBN 9781637746097 (hardcover)
ISBN 9781637746103 (electronic)

Editing by Leah Wilson
Copyediting by James Fraleigh
Proofreading by Jenny Bridges and Cheryl Beacham
Indexing by WordCo Indexing Services, Inc.
Text design and composition by Aaron Edmiston
Cover design by Pete Garceau
Cover images © iStock / bauhaus1000 (Gardenia and Camellia; Fringeflower, Lily and Narcissus) and NSA Digital Archive (Bishop Rose, Common Camellia, Yellow Roses, Duchess of Orleans Rose, and Hundred Leaved Rose)
Printed by Lake Book Manufacturing

**Special discounts for bulk sales are available.
Please contact bulkorders@benbellabooks.com.**

For Dad,
who showed me that
the frivolous and the serious
are better together than apart.

CONTENTS

Foreword—Daniel Z. Lieberman, MD | xiii

Introduction | 1
**LET'S FIND SOME ANSWERS, THEN
LET'S LOOK FOR . . . MORE**

PART I: UNDERSTANDING THE MOLECULE OF MORE

*To Begin: An Ancient System Straining
Against a Modern World* | 13

Chapter 1 | 15
HAPPINESS VERSUS SURVIVAL

PART II: TAMING THE MOLECULE OF MORE

To Begin: Going Up? Going Down? | 37

Chapter 2 | 39
BOOSTING DOPAMINE

Chapter 3 | 53
THE DOPAMINE FAST

Chapter 4 | 65
DOPAMINE TOOLS THAT MIGHT BE COMING SOON

Chapter 5 | 81
RAISING AND LOWERING DOPAMINE THROUGH THERAPY

PART III: TAMING BIG ISSUES, ONE AT A TIME

To Begin: How to Become a Lion Tamer | 103

Chapter 6 | 105
WRESTLING WITH ROMANCE

Chapter 7 | 113
FINDING THE RIGHT PERSON

Chapter 8 | 127
MANAGING SEXUAL FEELINGS

Chapter 9 | 135
HOW TO STAY TOGETHER

Chapter 10 | 149
PRODUCTIVITY AND GETTING AHEAD

Chapter 11 | 163
TAMING SOCIAL MEDIA

Chapter 12 | 175
OBSESSING OVER THE NEWS

Chapter 13 | 181
ONLINE PORNOGRAPHY

Chapter 14 | 191
BUYING STUFF

Chapter 15 | 199
GAMING

Chapter 16 | 209
CULTIVATING CREATIVITY

FINALE: THE NEED FOR MORE THAN MORE

In Conclusion: Finding a Way to Live | 225

The Sound of a Tuning Fork, Struck Upon a Star | 227

Author's Note | 241

Notes | 243

Index | 253

FOREWORD

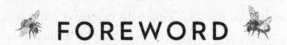

Daniel Z. Lieberman, MD
Co-author of *The Molecule of More*

What's the takeaway? That question echoed in our ears following the release of *The Molecule of More*, a book that unexpectedly became an international bestseller. The profound impact of dopamine on our lives had struck a chord with readers worldwide, but they craved more—a guide to harnessing this powerful molecule's potential.

The best we could do was to say, "The takeaway is knowledge." That answer wasn't what our agents or publisher wanted to hear, but it was enough to set off a global conversation.

As the years progressed, the calls for a second book grew louder. Readers needed more than understanding; they wanted strategies to navigate the dopamine-driven aspects of their lives. So, Mike and I sat down to discuss writing a sequel. Our professional lives had evolved since our work together on *The Molecule of More*, and we decided to approach the sequel from a new direction that would make the best use of our individual strengths. I did much of the research for the book, combing through studies that would help us better understand how to manage the complexities of dopamine's influence over our lives. We

FOREWORD

worked together for many months on early drafts of several chapters. But when it came to turning that early work into a book, Mike took the helm.

As a seasoned writer across multiple genres and a respected educator on the faculty of Georgetown University, Mike possesses the unique ability to illuminate intricate subjects with a storyteller's flair. His background in the hard sciences, along with extensive experience in writing screenplays, novels, and memoirs, gives him an ability to translate complex concepts into relatable narratives that's nothing short of artistry. This is an essential skill when it comes to something as complex as learning how to transform the untamed energy of dopamine into a powerful ally.

Mike's sequel boldly addresses the challenge of dopamine's dual nature—a source of inspirational genius but also a siren call to our baser instincts. It's a journey through the most promising strategies for balancing dopamine's scales, favoring fulfillment over folly.

This book doesn't offer easy answers because easy answers usually fail when confronted with life's complexities. Instead, it presents the hard answers—those that reveal the most fruitful avenues for your energy and perseverance. Dopamine is both powerful and clever. Cultivating strength of character is necessary to make it work for you, but it's not enough. You also need to understand dopamine's tricks—and how to outsmart them.

I hope this sequel will teach you how to control dopamine, so dopamine doesn't control you. It will show you what to pay attention to, when to persevere, and ultimately, how to arrive at a life well lived. Prepare to be guided through the labyrinth of the mind to discover the secrets to gaining control over the hidden power of the molecule of more.

There are two tragedies in life. One is not to get your heart's desire. The other is to get it.

—George Bernard Shaw, *Man and Superman*

Introduction

LET'S FIND SOME ANSWERS, THEN LET'S LOOK FOR . . . MORE

The most important chemical in the human brain is dopamine. That's not to say that other neurotransmitters are trivial. In fact, they are vital. Serotonin is often central to depression and its treatment. Endorphin affects mood and helps mitigate pain. Oxytocin is essential to sex, attachment, and maternal behavior. Anxiety, eating disorders, dealing with trauma—to address them, we would need a sophisticated understanding of many brain chemicals at the same time.

But when it comes to the moment-to-moment experience of modern life, dopamine is where we begin. It is the key to understanding the urges we feel and why we respond the way we do. In most matters, if you want to improve your outlook and behavior, stroll past the other neurotransmitters and start here. It isn't even close. Navigating a romantic relationship? *Dopamine.* Separating lust from caring? *Dopamine.*

Seeing through the attraction of risky behavior? *Dopamine*. Taming the lure of online porn, infidelity, social media, doomscrolling, getting high, giving up, putting work over family, or pushing your most important responsibilities a little further down the list of things to do? *Dopamine, dopamine, dopamine.*

When it comes to navigating any choice in life, dopamine pushes us toward acquisition and discovery. That's what it evolved to do because those are useful things—they keep us alive. But the nature of the world means that along that same path, we'll also find damaging, even dangerous options—troubling temptations and bad ideas with visceral appeal. At those points, dopamine can and will lead us astray. Here's why: It's telling us at every turn, *How about you trade what you already have for the possibility of something even better?*

It's just ahead, says dopamine. *Trust me.*

Until the past few hundred years, the primary challenge humans faced was staying alive. Anything new and unknown could be dangerous—or useful, even lifesaving. With threats and opportunities around every corner, the dopamine system evolved to give us an instant, profound, and insistent urge to investigate anything unfamiliar and unexpected. In addition, dopamine evolved to match that urge with an on-the-fly capacity for planning, reasoning, and creative thinking when we need it most. Thus, not only are we fascinated by the unknown, we also leap to explore it. Dopamine's one-two punch has led us forward as a species: it drives us to acquire an ever-growing understanding of the world around us, then to apply what we've learned to gain dominion over it all.

Fast-forward to the twenty-first century, where we stand astride the world as gods. Cell phones, televisions, microwave ovens, air travel, even indoor plumbing—we have mastered our home. The science fiction of even recent decades has failed to predict as spectacular a way of life as what we have achieved. A monarch in Shakespeare's time would tremble at a tour of our tiniest grocery store. (If he or she got that far. The sovereign would've had a heart attack at the sight of the car we'd take to get there, never mind the sorcery of electric lights and paved roads.) Yet we never experience a day without at least one moment when we decide that this or that miracle is getting just a tad stale—spouse, job, car, apartment, TV, sneaker, waistline, bank account. Enter dopamine.

Surely I can do better, we think. *I want . . . more.*

Thus the irony: Dopamine has enabled us to build a world that answers most of our needs, even our whims, and has rendered that world nearly devoid of immediate threat to the point that today we rarely need dopamine's hair-trigger response and hardcore intensity. And since evolution is slow, this leaves us with a new problem: much of what dopamine does is now not only unnecessary but also problematic. Once they had served their purpose, dopamine's more forceful powers didn't make a timely evolutionary exit. They stuck around. In the modern world, that means dopamine is frequently a villain.

Dopamine does most of its vital functioning outside of our awareness and stirs up trouble wherever our conscious mind is focused. We now catch dopamine asserting itself in situations where it has nothing constructive to contribute. It revs our anticipation, drives us to distraction over unlikely threats, and tempts us with dubious opportunities that borrow our peace and break our hearts. How odd that the biological system that equipped us to build this world full of miracles would now so frequently reduce us to twitchy, dissatisfied subjects of the domain it helped us create.

Seven hundred years ago, Chaucer wrote that familiarity breeds contempt, but he identified only half the problem. Familiarity also breeds boredom, and boredom is shorthand for the experience of lacking a challenge. That matters because humans are in constant need of at least a little pushback, and that's why the biological imperative driven by dopamine will assert itself no matter how amazing, desirable, and safe our environment happens to be. This part of the brain has wired us for restlessness. As soon as our little corner of the world runs out of surprises to keep us amused, we get antsy. That's when the molecule of more sends us spinning off in search of something new, a restless casting about for amusement often masquerading as genuine need.

This problem isn't new. It's built into the human condition. Adam and Eve had the entire Garden of Eden to enjoy and even they got restless. Once you cross the finish line and claim the prize you sought, it's only a matter of time before you get bored with it.

Not that we're helpless against this. Dopamine drives us toward more, but it's up to us to decide whether to embrace that motivation, modulate it, or turn it down.

That's hard. We need to tame dopamine when it pushes us toward the *more* we don't need, and let it do its thing when it promises a *more* worth having. This skill does not come naturally. Building our capacity for it is the subject of this book.

We'll need something else too. We'll need something to make us remember at every turn that this problem is more than biology and chemistry.

Against the push of dopamine-driven demands, we will need to find an outlook to give us a meaningful way to navigate life. That's a tough one, and it has troubled humans since we first appeared on the earth. We have to figure out how to live a life that gives us not only peace for today but also optimism for tomorrow, and we'll have to do it in a world that seems inclined against both. It's a tall order made more demanding by the fact that the solution will not come falling from the sky. We have to find it.

In the film *Crimes and Misdemeanors*, here's how a character played by psychoanalyst Martin Bergmann memorably described the problem:

> *We define ourselves by the choices we have made. We are in fact the sum total of our choices. Events unfold so unpredictably, so unfairly, human happiness does not seem to have been included in the design of creation. It is only we, with our capacity to love, that give meaning to the indifferent universe. And yet most human beings seem to have the ability to keep trying, and even to find joy from simple things like their family, their work, and from the hope that future generations might understand more.*[1]

LET'S FIND SOME ANSWERS, THEN LET'S LOOK FOR . . . MORE

Human happiness does not seem to have been included in the design of creation.

Despite the swarming anticipation that dopamine stirs, no permanent satisfaction is automatically coming soon, no sense of completeness is for-sure bound up in events just ahead, and no fulfillment is guaranteed to be extracted from chasing the river of maybe/maybe-not promises that modern life squeezes out of dopamine.

It is only we, with our capacity to love, who give meaning to the indifferent universe.

And there it is. Not only is the day-to-day challenge of dopamine ours to address but so is the deeper need for *meaning*.

By dint of human nature, we are bound to try to calm the bitchy back-and-forth inside us that lurches between anticipation and disappointment. This struggle appears to be an unavoidable byproduct of the design of the brain—the wonderful, miserable motor that drives our pursuit of progress. To feel better in the face of the day-to-day challenge of dopamine, we need coping mechanisms that reflect the facts of neuroscience. But to find meaning, we must craft a narrative to attach us to purpose across a lifetime.

I submit that the daily struggles and the need for meaning are bound up together. The solution is for us to do what evolution has not: find the *deus ex machina* or put one there ourselves—and not just any *deus* will do. We must find a meaning—or, more precisely, a pursuit of meaning—that helps us really cope simultaneously with the daily, even hourly, struggles that come to us from dopamine.

Let's begin here. The secret of life, as James Taylor sang, is enjoying the passage of time. Or as a wise young man once put it, "Life moves pretty fast. If you don't stop and look around once in a while, you could miss it."[2] We must decide to quit chasing the promise of tomorrow as if the pleasure we can take from today is a lesser concern. We must choose to take joy not just from pursuing what might be but also from appreciating what is. In this book, we'll collect ways to tame the bit of biology that links serial progress with nagging dissatisfaction, but we won't stop there. Our struggles and our triumphs ought to add up to something greater. A life that feels well lived requires us to balance the burden of wanting versus having.

It's ironic. We are overwhelmed by the urge to pursue every potential acquisition that appears before us. But to find relief from that, you and I are obligated to search for something . . . *more*.

This book begins, in part I, with an explanation of what dopamine does and how it works. A tour of this neurotransmitter and its brain systems will give you an understanding of the science behind much of the day-to-day human experience. Later in the book, when we discuss ways to help you deal with your dopamine-related challenges, this information will help you identify and more effectively apply the approaches we'll explore.

Part II describes things you can use to drive more or less dopamine activity overall. You'll find approaches to raise or lower the overall impact of dopamine on mood, urges, and behavior. There are compounds, treatments, and activities available right now that can make a difference in your day. We'll also explore other prospects now being investigated.

Part III provides support, one issue at a time, for specific dopamine-related problems. Here we explore step-by-step techniques to more effectively manage particular challenges driven by the molecule of more. These approaches will help you gain greater control over the feelings and choices that trouble you the most.

"Finale" describes the most critical element in dealing with dopamine. Having addressed particular dopamine-related problems and their solutions, we now have another opportunity: to find a way to deal with the foundational feelings that make those problems occur so frequently and so intensely in our lives. For this we will fuse the discipline of neuroscience with something just as significant from the realm of ideas. I'll propose a modern twist on an ancient way of navigating daily life so we might make for ourselves a more satisfying, more complete life.

What you find here is going to surprise you—in a good way. You *can* feel better. It starts with your decision to make the effort—and it's worth it. Read on.

WHAT CAN I DO RIGHT NOW?

The point of the first book, *The Molecule of More*, was to show you the surprising and profound influence of dopamine in life and culture. The point of *this* book is to equip you with skills to help you change that influence for the better in your daily life. The recommendations you'll find come from research into published, peer-reviewed medical and psychological studies. In some cases I spoke extensively with psychiatrist Dan Lieberman, my co-author on our original book, whose career-long specialty has been providing people with practical ways to help them feel better.

At this point, you may not be able to clearly identify which of your issues are related to dopamine, but would like to start making progress anyway, right away. Here are five simple things you can start doing now to improve the best effects of dopamine and reduce those that may be causing you trouble. Doing these things will raise your confidence and lower your anxiety—and give you a head start on taming the molecule of more.

Get regular physical activity. One well-proven way to increase constructive dopamine activity is to get some exercise. Beyond potentially boosting dopamine, regular physical activity also reinforces the exercise habit (via the dopamine reward system, which we will discuss in part I). The improvements that come with consistent exercise are likely to give you an improved mood, superior self-control, and a stronger sense of your own worth.

Prioritize sleep. Simply getting enough sleep, and getting it regularly, has far-reaching effects on the dopamine system. Sleep deprivation appears to decrease the positive effects of dopamine by reducing the availability of the receptors that allow it to function—think of having lots of cars in a city and not enough roads to drive them on.³ Good sleep habits help fix that. In addition, the dopamine system is tied to our circadian rhythms, the twenty-four-hour cycles of the body and brain. Disrupting one system can disrupt the other. Thus, good sleep provides bank-shot support for dopamine function, which in turn influences circadian rhythms related to vision, the sense of smell, certain types of motor function and reward systems, and general equilibrium among bodily systems.⁴

Put on some tunes. When you listen to music that you enjoy—and enjoyment appears to be key to the effect—several dopamine-oriented reward pathways in the brain are activated (including, for reasons not yet clear, pathways associated with survival). In addition, complex music with frequent variations in style, elements, and volume tends to create more neural activity than simpler music.⁵ Thus we have an easy way to stimulate the good feelings associated with dopamine: put on some tunes that really speak to you, songs that you love.*

Practice being still. Particular kinds of self-focus can stimulate dopamine. Consider yoga nidra, which is kind of meditation, but kind of not. Yoga nidra is best understood as the state between being awake and being asleep, a form of deep concentration and awareness that you can bring on yourself. It isn't necessary to engage with the spiritual and philosophical aspects of yoga to make use of the practice,

---

* Why do we like music in the first place? Part of it may come from surprise. Early in life, we learn how musical scales sound, and we recognize tension and release when the musician moves between chords. Music we like plays on these expectations. Since dopamine gives us a good feeling when we are surprised in a positive way, perhaps part of the attraction is that music often meets our ingrained expectations and then exceeds them through a surprising and enjoyable change of chord, key, or melody.

either. You can do it for its effects alone. But you'll probably need help. An internet search will lead you most often to lists of vague instructions like "visualize your goal" and "find a safe place in your mind"—not much help, that. Instead, try "guided meditation": find someone, either in person or on YouTube, who has done yoga nidra successfully and is experienced in talking others through achieving the state.

At least two studies indicate that yoga nidra (and yoga generally) increases our dopamine activity. Such a boost is usually associated with more engagement with the world around us, but the increase associated with yoga nidra seems to cause less engagement and activity, and that's a good thing. This decline gives us some relief from the demands of executive function, the part of the mind that helps us plan, focus, remember, and multitask.[6] Other benefits that yoga nidra provides include increases in serotonin, which helps govern mood, and in brain-derived neurotrophic factor (BDNF), which plays a role in several areas including long-term memory.

Eat right. Good nutrition keeps the body in balance and helps every system to function properly. That includes the dopamine system. A healthy diet will support your body's ability to produce dopamine; a less-than-healthy diet will undermine that ability. It's as simple as that.

Don't fall for dubious promises about "dopamine-enhancing" foods. Boosting your dopamine by eating or drinking some "miracle" food is about as likely as winning a Michelin star by tossing a little saffron into your boiled chicken. Unless you're severely malnourished, your dopamine problem is not due to a lack of foods containing the chemical components of dopamine. Lean meats, dairy products, and leafy vegetables can be good for you but they won't cause your body to produce extra dopamine any more than having flour, butter, and sugar in the pantry automatically leads to cookies in the cookie jar. The closest we have to dopamine-related foods are fruits, vegetables, and whole grains, which stabilize blood glucose levels. High glucose levels can unnecessarily stimulate the brain's reward systems, but consistent levels can help us avoid difficult bursts of urges, and for now, when it comes to dopamine and eating, that's about it.

Part I

UNDERSTANDING THE MOLECULE OF MORE

To Begin
AN ANCIENT SYSTEM STRAINING AGAINST A MODERN WORLD

This book is a step-by-step, topic-by-topic guide on how to tame the effects of dopamine—to enhance its impact when we lack motivation, to decrease it when we're overwhelmed, and to modulate it when we face a more nuanced challenge. Follow these techniques and you'll have powerful tools to improve your life. But before we try to change anything about your relationship to dopamine, let's start with a refresher on its basics.

Dopamine plays a crucial role in many of our conscious actions and reactions, particularly those involving motivation, reward, and choice. The feelings we have and the actions we take most often begin like this: *See that thing you don't have? That thing that's new to you? It's filled with possibilities and you should check it out. You want it. You may even need it. And if you make it yours, you'll almost surely be safer, more satisfied, more secure—maybe all those things. Don't doubt the feeling. Get busy!*

The problem is this makes us more susceptible to impulse and less inclined toward careful decision-making. We're driven more by allure than evidence, focusing on potential rather than reality. In the world of early humankind, where survival required minute-to-minute awareness, this constant sensitivity to potential risk and reward was ideal. But in a miraculous modern world filled with fantastic distractions? Such a system pulls us in a thousand directions, many of them destructive. Some attractions have genuine value, but the rest range from harmless diversions to deadly disruptions. What we want is frequently in conflict with what we need. The most common result of this conflict—probably what brought you to this book and its predecessor—is that this parade of attractions, some dubious and some desirable, often leaves us with ennui, dissatisfaction, and the sense that this constant desire for more is leaving us with less.

Though the provocations the dopamine system encounters today are vastly different from those that early humans faced, its response is still the same, and therein lies the problem: the dopamine system evolved for a life humans no longer live. It is up to us to adapt our responses. To do that we'll need a modern understanding of the ancient dopamine system. In this section, we'll review how it works, its two key parts, how we experience them, and the benefits and challenges this system provides.

Chapter 1
HAPPINESS VERSUS SURVIVAL

THE PROBLEM: CHOOSE HAPPINESS NOW ... OR BETTER THINGS LATER

What it takes to survive in the long run is rarely what makes us happy right now.

If I go to work, I can pay the rent ... but I'd rather sleep in.

This steamed broccoli will be good for me ... but, c'mon! This place has ice cream!

I ought to save some of my paycheck for a rainy day ... but I could buy this new TV right now.

Most of our moment-to-moment problems are like that: Do I want long-term reward or short-term fun? The things that enable our survival over time are rarely the things that make us happy right now, so we compromise. As individuals, we stick to a diet but give ourselves a cheat day. As a society, we agree to work Monday through Friday but take the weekend off. This sometimes casual, sometimes grave struggle between now and later—between discipline and delight—is a central feature of human experience. Sometimes there seems to be no middle ground at all: you can have a short, happy life or a long one built on self-denial.

But if you're willing to dig, there's some good news in all this: while unhappiness is built into the human brain, being unhappy has value. It's one of the greatest motivators there is. It doesn't take much to push us to act. For most people, all it takes is the tiniest sense that there is something better ahead, something more to be had, just beyond where we are right now. This is how we make progress in life. We are often willing to sacrifice a little of our comfort in the moment for the possibility—not the certainty, just the tiniest sliver of "maybe" (and that's going to be important)—that we can swap it for something better, something more.

Each decision we make is a tradeoff between today's happiness and tomorrow's "what if," and the sacrifice of one or the other is entirely up to us. In matters great and small, this determines how we feel and what we achieve. It is the engine of progress.

THE BOTTOM LINE

We must regularly choose between gratification today and value tomorrow. This often leaves us unhappy, but there's worth in that: being dissatisfied motivates us to create better things as individuals, families, organizations, and societies. Happy people don't improve themselves or the world. That might not sound like good news, but it's a powerful truth we can use.

THE SOURCE OF DESPAIR

The natural conflict between benefit tomorrow and pleasure today can leave us not just unhappy but in despair. As satisfaction from our victories fades, life often brings us back to the blues.

You got the job you wanted but in a while you feel disappointed because it wasn't what you hoped. So you start looking for the next job, and the cycle begins again.

HAPPINESS VERSUS SURVIVAL

That designer dress you bought looked great in the dressing room but after a party or two it no longer seems so special. You put it on Poshmark and keep an eye out for the next one.

If only he'd go out with me. If only she'd sleep with me. If only this one would marry me.

Not long after we succeed, a nagging question always returns: Is that all there is?

After enough times on that merry-go-round, many of us start looking for something beyond the chase for possessions and achievements, something to give our lives more significance. One approach is to put work over everything else—but then family usually suffers, and we often find our career isn't as fulfilling as we'd hoped. Or we try the opposite approach, putting family above work, but our financial security often suffers, or we miss the ambitious feelings from working so hard, or the kids grow up and move away and we end up alone with the spouse and the dog and *now what?* Other times we start living for hobbies, or social media, or travel, or volunteering, or education, or politics, or church, or gardening, or assuming the role of resident asshole at the homeowner's association, and those don't work, either. It really can feel like we're back on that merry-go-round—one in the most pointless amusement park in the world. We find ourselves traveling the same sad circle again and again, marked at best by disappointment, and at worst by depression, anxiety, troubling behavior toward others, troubling behavior toward ourselves, or some combination of those things.

🐝 THE BOTTOM LINE 🐝

We desire something, we chase it, and finally we possess it. That's how we get ahead in life. But achievement gives us only fleeting pleasure. Thus what pushes us toward progress can also drive us to despair. In this way a basic problem of human existence comes into focus: success and dissatisfaction are linked, perhaps inextricably.

AN ANCIENT SOLUTION OUT OF PLACE IN A MODERN WORLD

Early humans lived in a world of hair-trigger threats and scarcity. They needed to be naturally attuned to every unfamiliar change around them because anything new might hurt them, be their next meal, or prove useful in some as-yet unimagined way. Thus our behavior, via brain systems, evolved to ensure our survival.

But the world has changed. If life still consisted primarily of bodily threats and meals you need to hunt down the same day you consume them, this constant attraction to *more* would be ideal, but that environment is long gone. The twenty-first-century world is a pretty safe place. We don't have to engage with every prospective opportunity because, by and large, we are going to be okay. Thus, what was an evolutionary advantage is now in many ways a burden, one unique to modernity. Our survival no longer depends on a rabid urge to check out every minor mystery that crosses our path.

Yet we remain powerfully, biologically intrigued by every provocation and every possibility, so much so that the drawbacks of this hair-trigger system often interfere with normal life. If your next meal is not a worry, and you have a roof over your head and a coat in your closet, the urge to explore whatever is new is an urge too frequent and intense to serve you well. In the modern world, we end up chasing a lot of things we don't need—and suffering the fallout.

> ### THE BOTTOM LINE
>
> We evolved a drive to be curious about new things because they might help us survive. That was useful for early humans, but in the modern world it leaves us susceptible to unnecessary, sometimes challenging or even dangerous, distraction. This ancient approach is too sweeping for how we live now. In a world as safe and secure as ours, frequent, superfluous curiosity comes with a cost.

HAPPINESS VERSUS SURVIVAL

THE NEUROTRANSMITTER BEHIND IT ALL: DOPAMINE

The part of the brain responsible for all this is powered by a neurotransmitter (or brain chemical) known as dopamine. Dopamine and its associated circuits identify the possibility of something useful, then create a tantalizing feeling of anticipation and optimism that drives us to investigate. But note the illogic: the feeling is powered not by evidence but by the unexplored potential for value. Thus we're undertaking a lot of effort on untested optimism. Dopamine urges us forward not on the basis of *Let's weigh the pros and cons*, but *You bet it's good! Let's go get it!*

In other words, dopamine urges us to race after what *might* be good on the false promise that it *is*.

That's how dopamine can lead us to sacrifice the good things we already have for things that only *might* be good. It points beyond the horizon, where no one can see. Dopamine creates the itch that demands a scratch and drives us to take a chance, sometimes beyond our better judgment or good sense. Never for a sure thing and sometimes at a high cost, dopamine propels us forward to maybe, *maybe* make our lives better—more secure, more prosperous, more interesting, more everything. Dr. Dan Lieberman and I dubbed it "the molecule of more" for good reason. No matter how wonderful the thing we already have is, dopamine makes us dissatisfied with it and hungry for the next thing, no matter how uncertain it might be. This is why dopamine is so important to understand and tame.

Rainn Wilson, who played Dwight Schrute on the iconic series *The Office*, is famous, wealthy, and beloved by millions, yet he says that at the height of his success he felt miserable. He had a nagging feeling that, even with all he had, he ought to have more. Here's how he put it in a conversation with comedian Bill Maher (emphasis added):

> When I was in *The Office*, I spent several years really mostly unhappy because **it wasn't enough**. *I'm realizing now, like, I'm on a hit show, Emmy nominated every year, making lots of money . . . People love it.* **I wasn't enjoying it. I was thinking about, "Why am I**

not a movie star? *Why am I not the next Jack Black or the next Will Ferrell? How come I can't have a movie career? [. . .] It was never enough."*[7]

It's a classic dopamine experience. Whatever satisfying things we already possess, dopamine points beyond the here and now to say, *This thing you don't yet have? This is the real key to happiness. Go get it!*

Dopamine doesn't "know" anything certain about the promise it's making. Think of a guy standing in front of a nightclub trying to draw people into the comedy show just inside. He promises you the funniest night of your life, yet that guy has nothing to do with the show. He might not have even seen it! He's a salesman with a single job: to get people in the door. Will you really enjoy the show? He's gonna tell you that you will, but he doesn't know for sure.

That salesman is dopamine.

Dopamine is promising you a good time because that's what dopamine does. It's out there writing checks it'll never have to cash. Making good on those promises is somebody else's department.

🐝 THE BOTTOM LINE 🐝

Dopamine drives us to pursue whatever is new and potentially useful. The problem? It's triggered by possibility, not certainty—optimism, not evidence. This causes us to chase things that often have no value at all or that may even break our hearts.

DOPAMINE'S PARTNER: THE HERE AND NOW CHEMICALS

It's important to understand that dopamine is half of a key partnership in our consciousness. Dopamine is the "wanting" brain chemical.

HAPPINESS VERSUS SURVIVAL

It deals with things beyond our immediate control—the extrapersonal space—and does so via anticipation and planning.

Dopamine's complement is literally all the other neurotransmitters. We'll call them the "here & now" or "H&N" chemicals because they deal with things in the peripersonal space—the area that contains things within our physical grasp right now. As opposed to the anticipatory pleasures of dopamine, the H&N chemicals give us the enjoyment of seeing, hearing, tasting, touching, and smelling, which are consummatory pleasures.

The next time you buy something you've been saving up for and anticipating, pay attention to your feelings as the transaction is completed. This is the moment when the dopamine side of the partnership hands off to the H&N side, when the feeling of wanting gives way to the experience of having. You will be able to feel the switchover. Future-oriented dopamine pulls back. Sensory-oriented H&N chemicals take over. Now that you possess the thing you sought, your feelings are determined by what it feels like to touch it, to see it, to hold it, to experience it. Your anticipation of what might be is replaced by the reality of what is. The pursuit is over, and the dopaminergic attraction that brought you to this point begins to pass. Its motivational purpose has been carried out. The feeling goes away. The only thing before you is reality—what you have in your possession.

This change in feelings—this departure from the realm of dopamine into the realm of the here and now—is, often, the beginning of an eventual letdown. It's also the beginning of understanding why we so often grow bored with whatever we already have.

🐝 THE BOTTOM LINE 🐝

Dopamine gives us the ability to deal with things that are distant in time and space—things that we have to work and plan for if we are to secure them. The complement of dopamine is the set of brain chemicals that deal with the present

> and the close-in—the here & now chemicals. These are associated with the experience of consuming and doing. Dopamine is all about wanting. The H&N chemicals are all about having.

AN ALARM SYSTEM PLUS A CALCULATION SYSTEM

The dopamine system is made up of two subsystems. One we can think of as an alarm system. The other is a processing tool for figuring out what to do about whatever set off the first subsystem.*

Consider the first system. When we feel the urge to pursue some new and mysterious opportunity, that's a mental alarm going off. This is the *desire dopamine system* in action. When we are weighing a decision to do something presented to us by that system, planning how to do it, gaming out possibilities, and imagining scenarios, we are using the second system, the *control dopamine system*. Control and desire operate as partners focused on the future, lashing together feeling (desire) and thinking (control). The former makes us want to move ahead. The latter gives us the ability to do so. We start with the raw material of an unrefined urge, then process it with sophisticated and sustained calculation.

To reach our goals we need both, but it is desire dopamine† that keeps us on the path. Desire dopamine "promises" that we'll experience a profoundly good feeling when we reach the goal and delivers that motivation all along the way.

---

* There's a third dopamine system, but we won't address it here because it deals mostly with motor control. If you wonder why a Google search on "dopamine" yields so many hits for Parkinson's disease, that's why. Parkinson's primarily affects motion.
† A dopamine molecule causes one set of effects upon receptors in the desire system and a different set of effects on receptors in the control system. For convenience, let's refer to this distinction as different types or kinds of dopamine, as we did in the first book—for instance, "control dopamine" refers not to a unique kind of dopamine molecule but to "activity in the control dopamine system."

HAPPINESS VERSUS SURVIVAL

The desire circuit begins deep within the brain in the ventral tegmental area, where a great deal of dopamine is produced. It ends in the nucleus accumbens, which influences mood and behavior via the limbic system. When the desire circuit is activated, we experience wanting for things we don't have, plus related feelings such as motivation, enthusiasm, and anticipation.

The desire circuit also gives us a tantalizing feeling about what we're chasing. It's an intense experience that's difficult to describe. Think of it as the pleasurable feeling of possessing something . . . before you actually possess it. This in-between bit of emotional business is often superior to the experience of securing what you desire. Consider that moment in a restaurant when a server puts a spectacular meal in front of you. The meal is yours, but you haven't yet tasted it. This sensation combines the pleasure of anticipation and possession without being either one. Anticipation is mostly in the past, and possession is mostly in the future. They're both good feelings but they're definitely different. Which is better? Let's ask Winnie-the-Pooh:

> *"Well," said Pooh, "what I like best—" and then he had to stop and think. Because although Eating Honey was a very good thing to do, there was a moment just before you began to eat it which was better than when you were, but he didn't know what it was called.*[8]

In that moment "just before," desire dopamine is doing what it does best: motivating us with a tantalizing promise of a great here-and-now pleasure *in the future.* By giving us a taste—a version—of what might be possible, the future-focused desire circuit motivates us to overcome obstacles between ourselves and the goal.

This is the exact moment that dopamine can lead us into trouble. All that matters to the desire dopamine system is that a thing is new and novel. It "wants" that thing without considering whether it is healthy, wise, or responsible—there's no reason yet imposed, no rational deliberation. If the desire circuit was all we had—or if it was out of tune with the rest of the brain, or damaged—we'd chase after every new thing we came across.

Fortunately, desire dopamine has a partner, control dopamine, to introduce some thoughtful consideration. The control circuit begins in the same place as the desire circuit but follows a different path. Instead of releasing its dopamine in the limbic system, where we modulate basic emotions, it directs its dopamine to the frontal lobes. That's important because that's where planning and calculation take place. The combination of limbic-system motivation and frontal-lobe calculation gives us profound power. First, the desire circuit creates interest in what's possibly useful. Next, the control circuit lets us figure out if it is worth pursuing, and if so, how to go about pursuing it. It does so by supporting abstract thinking, providing focus, and giving us the ability to concentrate through complex mental tasks.

To reach our goals we need both subsystems working together.

> ### 🐝 THE BOTTOM LINE 🐝
>
> The desire dopamine system motivates pursuit toward new or novel things that may have existential value, and it does so on the basis of allure instead of fact. To secure those things, the control dopamine system equips us with the power to imagine, plan, and execute.

SPECTACULAR UPSIDE #1: MENTAL TIME TRAVEL

When dopamine presents us with the urge to pursue some goal, it doesn't make us feel desire, then leave us high and dry. Via control dopamine, it gives us the capacity to craft a plan to achieve the goal. For instance, it equips us to play out various scenarios: *What would it be like if I get what I want? How hard might this be? What challenges could get in my way, and how hard will it be to get around them?* In this way we gain a logic-driven tool to compete with the emotional, dopaminergic urge—a rational capacity to help us decide if we want to pursue the possibility at all. This combination of applied reason, intuition, and imagination gives us the power of *mental time travel*.

There's a restaurant near my house that makes a grilled chicken sandwich I like. I don't have one now, but I can play out various scenarios to help me decide whether to get one. *Do I want to spend the money? Would it be worth the trouble of driving over there? Or would I be just as happy staying home and microwaving leftovers?* And if I decide to have the sandwich, control dopamine gives me the power to make a plan for getting it.

My decision about that chicken sandwich is a trivial matter in the grand scheme, but the ability to carry out this kind of analysis is indispensable to navigating daily life. To choose my future, I must first be able to imagine my future. I must have the capacity to engage in mental time travel, meaning I need the ability to envision various future possibilities, analyze them, and tweak them to reflect benefits and problems, then game out those possibilities. Control dopamine makes that possible.

Consider a consequential decision such as proposing marriage. Desire dopamine is what piques my interest in formalizing a relationship, and control dopamine enables me to plan how I might make the case to my partner. But control dopamine also lets me make the decision of whether or not to ask. It lets me envision what my married life would be like down to the slightest detail, consider the benefits and challenges, and possibly realize that I'm not cut out for marriage at all, or that I'm perfect for it. Desire dopamine attracts us with possibility. Control dopamine gives us the power not only to plan but also to pursue, temper, or abandon a choice. It does so by allowing us to create imaginary alternative futures from which we can choose the best path.

Using mental time travel, I can plop myself down in multiple situations, play out the benefits and drawbacks, and make a choice about the future without investing anything but my mental energy. No risk. No danger. No cost. For instance, if I'm thinking about buying a car, I can imagine what it would be like to own an expensive SUV versus an inexpensive compact, gaming out issues of price, value, repair, utility, reliability, safety, and satisfaction. If I'm a college student, I can compare and contrast majors. If I'm a plumber, I can weigh the advantages of accessing a pipe through the tile inside the shower versus the drywall on the other side. As an engineer, I don't need to wait until a bridge is built to see if it can handle the weight of the cars. I can use the abstract laws of physics to calculate it in advance. As human beings, we can even think about entire regimes of abstraction such as justice, morality, and aesthetic beauty. Humans have more dopamine than any other species. The result is nothing less than the ever-improving world we live in—all because, before it was real, we crafted it as a dopamine-powered possibility made from pure thought.

The future begins with the desire for more than we have right now. This urge to look beyond, and our ability to imagine how to get a thing that's distant from us in time or space, equips us to make progress. "*Never enough,*" as Rainn Wilson put it, "has helped us as a species."

🐝 THE BOTTOM LINE 🐝

Dopamine comes with a spectacular benefit: it gives us control across time and distance, enabling us to imagine things that we can't yet possess or that do not yet exist. Dopamine also lets us manipulate abstractions and scenarios that exist only in the imagination. This skill, which we'll call mental time travel, separates humans from other species and is vital to progress and creativity.

SPECTACULAR UPSIDE #2: THE POWER OF A GOOD SURPRISE

The trigger for desire dopamine is not simply the opportunity to secure something of value. The thing that gets your attention must have the potential to make things better for you than you expect.

Let's say it's payday. My boss comes by my desk and hands me my weekly check. For the sake of easy arithmetic, we'll imagine it's $1,000.

HAPPINESS VERSUS SURVIVAL

I stick it in my wallet and go back to work. There's no dopamine buzz involved. But let's say that this week, my check includes a $100 bonus, and the note with the check says the boss has a new policy. From now on, she will sometimes add a bonus. Now I start looking forward to payday every week. But wait. It's only a hundred bucks, and it's not even a sure thing. Why was I not just as excited each week for the sure thing—which was ten times more than the bonus?

Dopamine is triggered by the potential to exceed our expectations in a good way—the power of a good surprise, or what we'll call *reward prediction error*. I thought I was going to get the same check as always. Instead, it was one with an extra $100. This exceeded my expectations. From now on, I know that possibility exists every Friday. This looking-forward-to-it feeling that comes from even the slightest chance of *more* triggers the dopamine buzz. As for why I'm excited by the extra hundred bucks when I yawn at the usual thousand, note that the trigger for the good feeling isn't the money itself. It's the unexpected thrill of an error in expectation that leaves us not with less or the same but *more*.

If your partner surprises you this Friday with flowers, it will trigger the good feeling of a dopamine release. How good? The dopamine release will be proportional to the difference between what you expected (your partner's arrival) and what actually happened (your partner plus flowers). But if he or she starts bringing home a bouquet every few weeks, your expectation of flowers will no longer be zero, so the dopamine release will be a little less each time. And if the one you love starts bringing home flowers every Friday, the dopamine release will return to zero because there's no difference between what happened and what you expected. The flowers used to be like a little lottery win. Now they've become something like that regular paycheck.

We'll return to this idea often. For instance, reward prediction error is why we keep scrolling through social media: *Maybe the next thing I read will make me feel better than I do right now. It doesn't happen every time but it's happened before. Maybe it'll happen again.* It's also the principle behind slot machines: *Maybe the next pull will put money in my pocket. It's happened before. Maybe it'll happen again.* Dopamine rewards us for pursuing even the most remote possibility of more.

> ### 🐝 THE BOTTOM LINE 🐝
>
> Dopamine is triggered by *reward prediction error*: when there is a possibility for something to exceed our expectations, we get a anticipatory feeling that is unique and intense. But once we grow accustomed to the surprise—when it stops being a surprise—the dopamine thrill goes away.

THE GREATEST DANGER: WHY IT TAKES MORE AND MORE TO FEEL LESS AND LESS

There's another characteristic of the dopamine system that's vital to know: the more frequently we stimulate any particular pathway in the dopamine system, the more stimulation is needed to get a response. Not only is dopamine on the lookout for "more," over time it requires *more* more. That can lead us to overindulge in things that can be very dangerous.

When you first started drinking alcohol in your teens or twenties, you probably felt tipsy after only a few sips. If you started to drink regularly, it soon took a cocktail or two to get that same feeling. Over months or years, and depending on how frequently you imbibed, you likely reached the point where it took many drinks to get what only a few sips once achieved.

In other words, it takes more and more to feel less and less.

Repeated journeys down a dopamine pathway produce this effect. The gratification fades a little each time until eventually there may be no gratification at all. Each experience raises our tolerance for stimulation: the more we indulge a particular dopamine reward, the less pleasure we get from it, so we end up reaching for more just to maintain it. This growing need for stimulation gets expressed in all sorts of adventurism, from mountain climbing to sex: every "next episode" has to be

HAPPINESS VERSUS SURVIVAL

even more extreme—more stimulating, more frequent, or both—to feel only as good the last time out.*

There's a better way. We can use our control dopamine system to moderate our choices, so that we don't stimulate those pleasure centers so frequently and so strongly as to render them impotent. Consider a case far less serious than illegal drugs, but with a similar pattern. A few years ago, I started drinking Diet Dr Pepper. I loved it and drank four or five cans a day, every day. Within a few months, though, I developed a complete distaste for it. I'd had too much, too often—too much stimulation and not enough time between indulgences to let my system return to a more normal, unstimulated state. If I had drunk less of it, and not had it every day, I might still be able to enjoy it. By choosing to pursue dopamine stimulation less often, we avoid "wearing out" the part of us that takes pleasure from it.

When we simply indulge dopamine for short-term pleasure with no regard for long-term damage—when we let the lure of desire dopamine overwhelm the judgment and planning of control dopamine—we can end up in endless pursuit. We chase satisfaction over and over, each time feeling it less and less, until we eventually arrive at this: *I'm always going to feel this way. This never feels as good as it used to. And I see no way out.* That's when we may end up with an overdose.

For the details, consider what's going on at the cellular level. A dopamine molecule brings about a feeling or urge when it encounters a dopamine receptor: the "lock" to the dopamine "key." You can use a metal key thousands of times in the metal lock on your front door and it'll always work the same way, but the dopamine lock is not nearly so resilient. If you stimulate the dopamine receptors too much and too often, the lock gets worn down. Some of those receptors simply stop working, leaving fewer places for the dopamine molecule to connect.

---

* People who abuse certain drugs sometimes experience this dopamine-driven experience in the most merciless way possible. A single use of certain drugs can in some cases warp the desire dopamine system so that the person's tolerance becomes infinite: no matter how much of the drug they take, no pleasure is possible at all. So they continue to take more and more, leading to overdose and death.

Fewer receptor interactions weaken the effect—the good feeling you get—when dopamine is released. Now it takes more stimulation to get the same effect as before. Your tolerance has increased in a process called *downregulation*.

But doesn't this point to a solution? *If the receptors stop working, the pleasure goes away. And if the pleasure goes away, the pursuit of that pleasure should go away too.* That would be true but for one other factor: dopamine produces not just pleasure but also craving, and craving does not downregulate and go away.

Consider someone addicted to cocaine. He started out using the drug because it made him high. After years of use he no longer gets the slightest pleasure from it. He has downregulated all his dopamine receptors into complete tolerance. Yet he keeps using the drug because even though the pleasure is gone, the dopamine-driven craving remains. The chemical source of craving, found in the nucleus accumbens, is the engine of addiction: pleasure fades but the urge does not.*

Craving is encoded within the brain in ways that are similar to how we encode memories, and we all know how persistent unpleasant memories can be. Overcoming craving can take years, often requiring the assistance of family, friends, healthcare professionals, and support groups. Anti-craving medications, such as naltrexone for alcohol dependence and buprenorphine for opioids, can also be powerful tools in the fight.† But it's difficult no matter what—doable, but difficult.

---

* Withdrawal is painful. It is the body attempting to return to normal function after its systems were warped by abuse. But withdrawal symptoms tend to fade. The primary engine of addiction is still craving.
† The new weight-loss drug semaglutide and related medications appear to diminish cravings of many kinds, not just for food. Later in the book, I'll explore what this might mean for a broad range of dopamine-driven problems, including addiction itself.

> ### 🐝 THE BOTTOM LINE 🐝
>
> When dopamine pathways are too frequently activated, we build up a tolerance to their stimulation. That's why we grow bored with some things and move on from them. But with certain kinds of overstimulation, especially from certain drugs and medications, we create an urge that doesn't go away. This is the mechanism of addiction: the pleasure fades or even stops but the craving remains, and we change our behavior to satisfy it, often in damaging ways. A few medicines can keep addiction from consuming us, but most often we have to find professional support to build coping mechanisms and address underlying causes.

THE BALANCE OF POWER AND THE POWER OF BALANCE

When it comes to creating a desirable future for ourselves, dopamine is vital as air, yet its power to make us despair—from too much anticipation or too little—is as common as a weapon at the scene of a crime. But with a basic understanding of the connection between dopamine and how it can make us feel, we are now better prepared to tame its undesirable effects.

On one end of the spectrum of dopamine experiences, there's the lack of desire. An overall deficiency of dopamine and a lack of activity in the desire system is often a significant factor in depression. It's crushing to feel that there's nothing you want and no one you care about. When even the desire to live is gone, a major depressive disorder can become a fatal illness. People who live with depression are twenty-five times more likely to take their own lives than the general population.[9]

At the other extreme, overwhelming desire pushes us toward constant, exhausting pursuit. Manic episodes are characterized by increased energy, restlessness, impulsivity, and the drive to engage in more activities than are practical. Among the main causes are too much dopamine, a too-busy desire dopamine system, and, in some cases, an overactive control dopamine system as well. There's little or no interest in here-and-now pleasure, just a permanent state of striving, often accompanied by inflated self-esteem. A person with such a problem is consumed by unending effort toward the next achievement, a victory they won't bother to pause and enjoy even if they get there. Perpetual, joy-free pursuit thus can become another type of addiction.

With these extremes in mind—profound lack of motivation at one end of the spectrum and obsession with pursuit on the other—we can more easily recognize where our own experiences fall. Are you concerned you are growing bored with something—a job, a relationship, a string of projects started and never finished? Have you experienced a dip in curiosity, motivation, or anticipation? These may be related to a lack of dopamine or dopamine activity. Near the other extreme, we may find ourselves with an interest in alcohol or drugs that interferes with our lives, an interest that is not quite or not yet an addiction. Or we may have problems with things we do. Perhaps we eat obsessively, or we engage in unwanted behaviors related to sex, food, shopping, or gaming, and we want to cut down or stop. Any of these things may be related to having too much dopamine.

When we fail to find pleasure in the things we have in the here and now, we need to diminish our reliance on dopaminergic anticipation and explore more deeply what we can experience with our five senses.

When we fail to find pleasure in anticipation, we need to raise our ability to look forward and engage with our curiosity and the unknown.

We must balance where we invest our energy and attention: the possibilities of the future should be balanced by the pleasures of the present, and the anticipation of tomorrow should be balanced by the experiences of today.

Ideally, here's what we'd like to happen every time we reach a goal. We stop to take in the pleasure of our newly-won whatever in the here and now. We indulge our senses. We take pride in what we've achieved. In a while, we will cast about for our next goal, and having found it we will begin to pursue it. At that point our here-and-now enjoyment will slowly and reasonably give way to another round of dopamine-driven motivation and the pleasure of anticipation.

Building this balance between the dopamine system and the here-and-now neurotransmitters is a matter of weighing what we have against what we want, the pleasure of what is versus the possibility of what could be—or, as I've written before, creating a livable balance between wanting and having. In a full and satisfying life, we appreciate both. It is in asserting control over the dopaminergic urge that we find a satisfying way to live.

This balancing act determines nothing less than our capacity to find peace and fulfillment in life. We have to be mindful of the risks and rewards of our attraction to the future. At the same time, we must make a conscious, purposeful choice to appreciate the present, especially in the period just after we've reached a goal. It's not easy. Not only does dopamine make anticipation and planning an intoxicating experience, it also makes us overestimate how satisfying any achievement will be. Think of how you felt when you started your first job. You worked hard to earn it. You believed it would be the key to happiness. Yet months later you may have found yourself wondering, *What else is out there?* The job was probably just fine, but "just fine" is the enemy of the dopaminergic urge.

We aren't wired for "good enough." We're wired for *more*.

Let's investigate how to tame this demanding molecule for a better life.

Part II

TAMING THE MOLECULE OF MORE

To Begin

GOING UP?
GOING DOWN?

Part II is a guide to raising and lowering your overall dopamine levels. If you feel that your levels are problematically too low or too high, this section will give you specific techniques to help change that. If you feel your levels are likely normal but you'd like to raise or lower them to try for some particular improvement in feeling or performance, I'll address that too. (In part III, I'll provide detailed support for particular dopamine-related problems.) I'll also share a preview of dopamine-impacting treatments now in development. Finally, I'll provide a guide to therapeutic approaches to dopamine-related issues.

What does it mean to have low levels of dopamine? In broad terms, low dopamine presents as a lack of interest in new or typically intriguing things, and diminished interest in pursuing activities and answers that were formerly far more interesting. People with low dopamine often complain of a lack of motivation, an inability to concentrate, depression or "the blues," brain fog, and a reduced interest in sex. A common symptom of depression is the inability to experience pleasure; this is associated with reduced dopamine signaling.

What does it mean to have elevated levels of dopamine? Elevated dopamine presents as decreased impulse control, excess energy, and sometimes aggression. Symptoms vary with the person but can also include racing thoughts and an affinity for risk-taking to the point that it becomes a problem.* In the case of certain serious mental illnesses, some people may experience hallucinations, delusions, mania, and even psychosis.

If you're experiencing a cluster of the feelings or behaviors just described, the techniques in this part of the book may be a helpful place to begin your efforts. (If you're in acute distress, you believe your issues require professional support, or you're just more comfortable working with someone, call a psychiatrist or a psychologist.) Improvement can come not only as a result of a genuine change in dopamine levels and activity but also via the general enhancement of your mental state these practices can bring about. Either way is fine because the outcome is that you feel better.

---

* Elevated dopamine is often associated with anxiety, but that in fact can be caused by both too much dopamine and too little, depending on where in the brain the activity occurs. For instance, elevated dopamine in the cognitive control and executive function pathways is associated with increased anxiety, while too little dopamine can cause anxiety if it occurs in the reward circuits. Also, dopamine isn't the only factor in anxiety. Other things matter, too, such as genetics, environmental influence, personal experience, and dopamine's interaction with other neurotransmitters such as GABA and glutamate.

Chapter 2

BOOSTING DOPAMINE

CAN WE OVERCLOCK THE BRAIN?

If we want more dopamine, can we get it? Is there a pill? Is there a procedure?

For many, a healthy brain is not enough. No surprise there, since dopamine makes us want more of everything, often including dopamine itself. It's seductive, this prospect of firing up the brain to raise our performance past normal. After all, Lexapro lifts the mood of depressed people. Can it do the same for those of us who aren't depressed? Adderall improves the cognitive functioning of people with ADHD. Can the rest of us get that, too? In general, no. We can "turn the dial" from below normal up to normal, but we don't have much of a way to keep turning it, at least not for long and not without problems. "Normalizing" interventions are rarely capable of more than that.

A computer chip with a faulty clock will run slow or not at all. If we fix it, we'll see a substantial increase in performance—back to normal. But for a chip that's working the way it's supposed to, there's not much we can do to squeeze out greater performance. Some people overclock their computer processors to get a little more speed, but the difference is subtle and they risk frying the chip. The brain is not so different. Those who try to "overclock" it with psychotropic compounds meant for other purposes often pay a high price in the form of addiction, depression, anxiety, or some other mental disorder.

Yet the prospect of taking control of the brain and making ourselves smarter or more productive "beyond normal" is so tempting that many people ignore the risks and try anyway. The failures and tragedies are well documented and plentiful. Success stories are few and far between, and "good" outcomes are mostly in the eye of the beholder, such as with the use of nootropic* compounds and dietary supplements. Some people report improvement, though this may be the placebo effect. Much of the research into these uses has yet to establish clear and consistent cause and effect.

As the next section describes, the body fights hammer and tong against the addition of dopamine itself to the brain. What we have for now are treatments that increase the activity of dopamine within the brain, but only under certain conditions and only in certain systems. These are often dopamine agonists, meaning they improve function by increasing the activity at dopamine receptors. Even so, the approach is still subject to the "not much beyond normal" limitation.

WHY CAN'T I JUST GET A DOPAMINE SHOT?

In general, when we're short on some chemical in our body, we extract it from a human or animal source or whip it up in the lab, then take it in a shot or a pill. When diabetics need insulin, they take an injection.

---

* *Nootropic* comes from the Ancient Greek for "mind" (*noos*) and "turning" (*trope*). It refers to compounds that enhance learning and memory, and build resistance to disruption of learned behaviors, doing so without making you sleepy or high. If you haven't seen this word before, it's because it's fairly new. It didn't appear until 1972, in a medical research paper by a Romanian chemist named Corneliu E. Giurgea who, conveniently, also provided this definition in his paper.

BOOSTING DOPAMINE

When you need to strengthen your bones, you take vitamin D. Why not do the same with dopamine?

The brain lives behind an elaborate filter called the blood–brain barrier (BBB). The BBB isn't a structure like the heart or the kidneys. It's everywhere throughout the brain, formed by special cells that line its capillaries, and it's huge. An adult has about two square meters of skin; the area covered by the BBB is eight times that.[10] The purpose of the BBB is to protect the brain and the central nervous system from toxic substances in the blood by rejecting molecules with certain qualities. This is where dopamine enters the picture—or, rather, does not.

First, dopamine is a polar molecule, meaning it has regions of positive and negative charge. The BBB blocks highly polarized molecules. Second, dopamine is a large molecule, meaning it has a relatively high molecular weight.* The BBB blocks large molecules too. This is why we can't introduce dopamine directly into the brain; for now there's no other practical channel of access.† This tells us something else too. Since there's no way to bring dopamine across the BBB, all the dopamine we use in the brain has to be made there. We could inject it all day long and not fix the problem. It would just float around in the bloodstream and never end up where it needs to be.‡

When we talk about increasing dopamine levels, we most often mean increasing dopamine activity by causing the dopamine receptors in the brain to keep the molecule around longer than they normally

---

* Such molecules are typically biologics, meaning they come from other living organisms, and this may be a clue as to why the BBB blocks them. There could be something dangerous about swapping molecules among living things that we don't fully understand. After all, the BBB is a product of evolution, so this quality of the barrier developed as a response over time to something we may not have seen in modern medicine.

† We can get around the BBB by going through the spine, but it's risky and requires a highly controlled environment. Also, dopamine acts and recedes far too rapidly for this approach to help for very long.

‡ Dopamine *is* found in the bloodstream, but its function there appears unrelated to behavior. It is a local paracrine messenger, meaning that it sends a signal to cause changes in nearby cells.

would. Doctors do this through compounds known as dopamine reuptake inhibitors (DRIs). This term may sound familiar because it is similar to selective serotonin reuptake inhibitors (SSRIs), the most commonly used class of antidepressant and anti-anxiety drugs. SSRIs make serotonin stick around longer in its receptor: after the serotonin molecule makes its connection, the SSRI keeps it from being swept away from its receptor as quickly as it otherwise would. The goal is to reduce feelings of depression or anxiety and elevate mood. In the same way, DRIs cause dopamine molecules to stick around longer in their receptors, extending the associated dopamine effect.

At this time, our main way of making the brain create more dopamine is to take in more of the chemical precursor to dopamine, the compound levodopa, or L-dopa, prescribed in a pill or as a transdermal patch. This compound is not without significant side effects and, as a matter of mechanics, it's kind of a bank shot—but it works, and it helps people greatly. L-dopa can do something dopamine cannot: it can cross the BBB and enter the brain itself. Once in the nervous system, additional chemical processes convert it to dopamine, though it's the kind associated with motion receptors. That's why L-dopa is a first-line treatment for Parkinson's disease. It would be valuable to have an L-dopa-like molecule that could become active in the control or desire dopamine circuits, but so far we haven't found it.

WELLBUTRIN, ADDERALL, SUGAR PILLS, AND LIES

Let's consider the potential for a dopamine boost from some more familiar prescriptions.

One of the most successful antidepressants is Wellbutrin. It has a special place in the hearts of psychiatrists because it treats depression just as well as other antidepressants without causing one of their most common side effects, sexual dysfunction. Unlike those drugs, Wellbutrin doesn't relieve anxiety or boost stress resilience, but it comes with other advantages. For instance, it can increase a patient's engagement with the world and make it easier for them to enjoy ordinary pleasures, including sex. It can boost interest, motivation, and the ability to experience pleasure. And it won't lead to addiction. Wellbutrin also boosts dopamine, but it delivers this effect only for patients with below-normal dopamine performance. For those blessed not to suffer from depression, Wellbutrin will not boost motivation, interest, or pleasurable response to pleasurable experiences.

Next, consider Adderall, the brand name for the stimulant known as amphetamine. Adderall modifies the behavior of the dopamine desire and control circuits. If you have attention deficit/hyperactivity disorder (ADHD), these circuits function abnormally and Adderall can make a difference. Modifying the desire circuit helps reduce impulsivity. For the control circuit, the effect is improved focus, concentration, and executive functioning (organizing, planning, and prioritizing). In addition to these desired effects, Adderall can also cause euphoria, elevated energy, and increased motivation.

If you don't have a diagnosed attention disorder and you're taking Adderall just for the anticipated effects, however, you'll likely run into significant trouble. You'll very soon find that you can never get enough euphoria and mental energy, and there's no natural brake in place to stop you from getting more of it. We know what usually happens when we can have all of a thing we want with no limits. Fortunately, most pleasures come with built-in guardrails. Sex, for instance, is a natural form of gratification with natural limitations, such as finding someone who wants you in the first place. But there's no such limit on stimulants that come out of a lab, especially one that makes us feel like a million bucks. Increased energy, motivation, and euphoria last only a while. The brain adjusts its tolerance, and after a few days or weeks the effects go away, so we increase the dose. The most likely result is that we take more and more of the drug until we overdose.

Another problem with the mood-boosting effects of stimulants is that the way you feel can be very different from how others see you. One user described feeling "charming, witty, and clever," while his family described him as "arrogant, snide, and condescending."[11] Overclocking the dopamine system in this way can wreck our otherwise casual ability to interact with people in a reasonable way.

At this point, though, we encounter what makes Adderall so valuable for people who have a dopamine deficiency and actually need it. Among people with ADHD, and only among those people, the brain makes a weird and welcome exception: it doesn't develop tolerance for the drug and it doesn't downregulate the system, at least not for its desirable effects. Once a doctor finds the right dose, there's no need to keep increasing it. The patient can use it long term to enjoy normal levels of focus, concentration, and executive functioning. It's not clear why this happens but it does. Adderall is profoundly useful if you need it, and profoundly dangerous if you don't, as it can send a healthy brain sideways, potentially toward dangerous, even deadly outcomes: mania, psychosis, and death.

None of that has diminished the popularity of Adderall as an occasional brain booster. One survey found that about 1 in 12 students reported taking it for nonmedical use.[12] What might that use be? Here's a hint: Consumption tends to peak around final exams. A group of scientists conducted an analysis of campus wastewater* during final exam week, then compared the data to what they measured in wastewater in the first week of classes, and found the levels of amphetamine rose 760% during finals.[13] But these students aren't gaining the advantage they believe. Investigators administered a battery of neurocognitive tests to a group of student volunteers, eight hours per subject, twice for each person, once with Adderall and once with capsules containing an inert powder. The results were the same under both conditions.[14]

Unless you have ADHD, you don't get "smarter" using Adderall. Other studies have found similar results. Yet in a 2021 survey of surveys, as many as 34% of college students were using stimulant medication to try to boost their academic performance.[15]

So why do students keep at it? Because what matters is the outcome. They really are experiencing a boost in performance, but not for the reasons they think.

Behold, the power of a lie. If you give someone a placebo and tell them it's a stimulant, they'll perform better than if you give them

---

* Wastewater contains urine and Adderall can be detected in urine.

a stimulant and tell them it's a placebo. In other words, convincing yourself of your edge is more powerful than ingesting some chemical. Scientists call this a "robust" placebo effect, and it means just what the word "robust" suggests: it's strong, vigorous, and better than expected. Scientists don't know the method of action for why this happens. They know only that it does. A little more energy might help you at exam time, but there's no line we can draw between that and being genuinely "smarter." All we know for sure is that it's as effective to believe you're taking a stimulant as it is to actually use one. Safer too.

And so far on our search, we're still not finding a drug that boosts normal dopamine levels.

EAT THIS

Next, let's look beyond formally-approved pharmaceuticals to over-the-counter supplements. But first a little history.

For nearly two centuries, you could walk into any US pharmacy and buy any medicine you wanted. In 1938, the federal government took away that choice in the name of consumer protection, collecting a catalog of compounds into a category for which a doctor's permission was needed for purchase. (That list, later expanded into five "schedules," has since expanded further to include four hundred or more drugs, with more added every year.) This system prioritizes minimizing risk at the expense of consumer choice and speed-to-market. That makes most people safer. But for those who are suffering with unusual symptoms or maladies, it creates a potentially life-and-death problem. If they want to try a drug that is not yet tested to current medical standards, they can't. Mercifully, after a half century of this regime, and with so many new potential treatments being offered, Washington carved out some exceptions.

The passage of the Dietary Supplement Health and Education Act of 1994 allowed manufacturers to bring vitamins, herbs, botanicals, and oils to market without the extensive testing required of pharmaceutical agents. Maybe they will help and maybe they won't. Maybe the outcome depends on the person, or the particulars of the condition. Whatever the case, people were free to choose their own balance between risk and reward. That's a good thing. But this also opened the door wide to dubious claims and false hopes.

Nutritional supplements that claim to boost dopamine are plentiful, carrying names such as NeuroSpark, Dopamine Brain Food, and DopaRush Cocktail. They promise benefits such as renewed energy, elevated mood and focus, and even better memory. Do these alleged dopamine "supplements" work?

You can't boost dopamine by simply consuming dopamine because, as discussed, dopamine can't get through the blood-brain barrier. But the most common ingredient in these supplements, L-tyrosine, can. L-tyrosine is a precursor to L-dopa, a dopamine precursor. A common claim is that getting more of L-tyrosine could boost dopamine levels in your brain.

Can it? In a word, *no*. That's like saying if I eat some calcium, I may grow more teeth.

Just because the raw material is present doesn't guarantee that the body will use it the way you hope. It may not use it at all. Or the extra stuff may throw off other systems that were working just fine until you started honking around. The brain creates and regulates dopamine using many complex processes. It's as yet unproven that we can accelerate those processes simply by adding more raw material to the mix. You might as well try to make a bigger cake by adding a whole bag of flour.

When it comes to supplements, the shelves are filled with breathless promises about dopamine levels and dopamine-related issues. In 2009, the National Institutes of Health sponsored a large study to test whether *Ginkgo biloba* extract improves cognitive health in the elderly.[16] It didn't. Researchers tracked more than three thousand volunteers for an average of six years. *Ginkgo biloba* did no better than placebo. Other supplements that are marketed as boosting dopamine do seem to have some effect on mental performance but not because they boost dopamine. L-theanine, which is found in tea, appears to increase mental

BOOSTING DOPAMINE

alertness while relaxing the mind, but the dopamine connection is a maybe at best. Theanine* achieves its effects by modulating multiple neurotransmitters and receptors in complex ways we don't fully understand.[17] A lot of companies are slapping "Dopamine Booster" on their products on the basis of inconclusive studies and wishful thinking.

In defense of supplements, these products tend to have good safety records, and most people use them *as* supplements, not cures. 79% of American adults say they take at least one nutritional supplement.[18] Total supplement spending is more than $30 billion a year. And even if these products don't work inside the body the way they're advertised, they can still be useful. Remember the placebo effect? People who take supplements tend to be living a pretty healthy life anyway. If these supplements encourage them and don't cause problems, that's a good thing. People who report consuming nutritional supplements also tend to report that they are in excellent health, drink alcohol only moderately, don't smoke cigarettes, and exercise frequently.[19] These are health-conscious people who are trying to maximize their physical and mental abilities.

There's little harm in experimenting to see what most so-called dopamine supplements can do, but they're unlikely to produce a true dopamine boost or dramatically improve focus, concentration, motivation, or energy. There is no substitute for a compound tested in rigorous human trials, and nothing out there now is a valid replacement for tested medicines. When it comes to neuro-enhancement, begin with common sense: if it sounds too good to be true, it almost certainly is.

---

* The terms *theanine* and *L-theanine* are used interchangeably. The "L" indicates the biologically active form, meaning a chemical structure inclined to interact with living organisms. For you chemistry enthusiasts, the two molecules are isomers, meaning they have the same number and types of atoms but in a different structural arrangement.

PROMISING NOISES FROM YOUR STOMACH

Finally, let's consider one of the newest fields of research in this area. What benefits might lie in manipulating one of our most mysterious internal systems, the "gut biome"?

Your body is like a bustling metropolis and its hustling citizens are microbes—bacteria, viruses, and fungi. There are trillions of them covering your skin and colonizing your intestines. Some are harmful and can cause illness. Some are helpful, even essential, to healthy physiological functioning, and they do remarkable things. Microorganisms in the stomach, intestines, and colon comprise the "gut microbiome" or "gut biome," a natural ecosystem of microbes that help break down food, aid digestion, and more. They're involved in the production of essential vitamins such as vitamin K, for blood clotting and bone-building; and some of the B vitamins, which support the mechanics of metabolism and have to be replenished regularly, since the body can't store them. These microorganisms also support the immune system, thus protecting the body from pathogenic microorganisms with less noble goals. But the most interesting thing they do in terms of our interest is produce compounds that influence how the brain functions.

Most of us would be surprised to learn that the gut and the brain communicate, and not just about what we'd like for dinner. Some of these signals travel through a network of nerve cells. Others take a more surprising route: they are transmitted via psychoactive chemicals that travel from the gut to the brain via the bloodstream.

Researchers, drug developers, and healthcare providers are increasingly interested in whether mental illnesses can be treated by manipulating the gut biome. Early results are encouraging. Working with volunteers who had major depressive disorder, scientists treated them with a probiotic supplement that contained three strains of bacteria:[20] *Lactobacillus acidophilus*, commonly found in yogurt; *Lactobacillus casei*, used in making cheese; and *Bifidobacterium bifidum*, found in the human intestine and essential to life. Researchers introduced no directly psychoactive pharmaceuticals; this was an experiment with supplements alone in the form of bacteria. The results: Volunteers who took the probiotic had lower depression scores than those who didn't. Similar findings have been reported for anxiety. As with depression, anxiety is related to dopamine function, and these bacteria seem to alleviate some measure of both.

This points us to an interesting possibility: Could these bacteria be part of new treatments for people who don't respond to medications? And to return to our original interest, are there any strains of bacteria that might provide a dopamine boost to an otherwise healthy person?

Scientists at the University of Pennsylvania Perelman School of Medicine addressed this last question directly. They examined the exercise habits of mice, some of whom spent considerably more time than others on the exercise wheel. They first sequenced the animals' genomes and studied various metabolites* in their bloodstreams, but neither effort explained the difference.[21] It wasn't until they focused on the microbiome that answers began to emerge. To confirm that it was the microbiome that was driving the exercise, they took some of the high-performing mice and treated them with antibiotics to kill their gut bacteria. The results were dramatic: the amount of time spent on the exercise wheel dropped by 50%.

Next, they bred germ-free mice—that is, mice born without any microorganisms in their gut. These mice didn't show much interest in exercising until researchers did a switch by transferring microbiome samples from high-performing mice to the germ-free ones. Afterward, the low performers improved to the point where their output closely matched that of the donor mice. Using antibiotics that selectively eliminated only certain strains of bacteria, researchers eventually found two specific strains that were responsible for the exercise boost: *Eubacterium rectale* and *Coprococcus eutactus*. These bacteria were producing special metabolites called *fatty acid amines* that led to increased levels of dopamine during exercise, which boosted motivation—at least for exercising,

---

* This sentence contains a couple terms that get tossed around a lot but are not in the working vocabulary of many people outside a biology lab, so let's clear up things. A *genome* is the blueprint or DNA for a living creature, including biological "plans" for growth and development. A *metabolite* is a product of chemical processes in the body.

and at least among mice. Voilà—a dopamine boost for those who are otherwise normal!

How this might lead to benefits for the rest of us is yet to be determined, but let's consider a few things we might try right now. If microbiomes really do enhance dopaminergic motivation, we can improve them right away. Western diets foster poor microbiomes. A good way to start fixing that is to add fiber and fermented foods to your diet. Probiotic supplements might also produce some benefits. Even if they don't specifically include *E. rectale* and *C. eutactus*, they may still be helpful: administering the common probiotic *Lactobacillus casei* can increase levels of other beneficial strains, including *C. eutactus* and *E. rectale*.[22] They also inhibit growth of some harmful bacteria. Other bacterial strains may have as-yet-unknown benefits. Prebiotics, which consist of soluble fiber acting as a kind of fertilizer for the gut, also promote the growth of *C. eutactus* and *E. rectale*. If you're trying to boost your motivation to exercise, a couple of teaspoons of whole or ground chia seeds might be a helpful addition to your daily diet, since their prebiotic fiber affects the growth of some of the bacterial groups at issue.[23]

STILL LOOKING

We set out to find ways to boost dopamine in healthy people and came up empty but it was still a useful exercise. In this new realm of medical possibilities, it's good to know what's already possible versus not yet possible, and why. Equipped with this kind of knowledge, we're less likely to fall for marketing that makes hope feel like fact—which, ironically, is the same thing desire dopamine does.

We've also ended with a little optimism on where progress might come: the gut biome may someday give us a way to boost exercise motivation. That's not the dopamine boost we were looking for, but if it leads to a treatment that really does motivate exercise, the impact will be lifesaving for some and life extending for us all. Even a little regular exercise reduces the risk of heart disease, high blood pressure, stroke, diabetes, and many other deadly illnesses. In terms of the brain, exercise directly improves cognitive functioning and boosts mood. Since more than 75% of Americans don't get enough exercise, in part from lack of motivation, a gut biome treatment that provides that motivation would be worth billions in new business, billions more in savings on healthcare costs, and years of longer, more satisfying life for many people—all from restoring a bit of motivation toward a common, or lately uncommon, act. The long reach of dopamine produces effects in every part of life.

It would be great to find a pharmaceutical—or a food, or a dietary supplement, or an altered piece of DNA, or a bacterium, virus, or fungus—that we could swallow like an aspirin to boost dopamine levels and benefits beyond normal, but right now there isn't one. When we try to push standard function beyond its limits, the results can include depression, addiction, and anxiety. Often the result is no change at all. Certain pharmaceuticals can restore function to a faulty dopamine system, but those compounds typically boost availability of the dopamine that's already in a healthy brain without increasing how much dopamine is there.

If we ever achieve neuro-enhancement for dopamine, the first successes may come from manipulation of the gut biome. Then again, science is about looking in new directions, so maybe the first breakthrough will come out of the blue. However it happens, it will come to pass because dopamine is ultimately the source of our motivation in this—as in everything.

Chapter 3
THE DOPAMINE FAST

FIRST, IT'S NOT REALLY A FAST

You want to go on a dopamine fast? Congratulations, you're already doing it.

Recall that you don't acquire dopamine by ingesting it. Every bit of it you have is made inside your brain. You could drink a tall glass of dopamine and your body would prevent every molecule from reaching the structures where it could influence your mood and outlook. That internal dopamine factory you have is humming along all on its own, thanks, and it requires neither your attention nor your consent to carry out its many jobs.

Besides, while you may sometimes feel overwhelmed by dopamine-driven effects, you don't want to eliminate dopamine entirely. You can't. Your body wouldn't function without it. When an infection co-opts your circulatory system, you don't suggest getting rid of your blood for a while.

"That's not what I mean," you say. "I want to temporarily eliminate dopamine-boosting *experiences*." Okay, but pretty much everything is a dopaminergic experience of some kind. Other than knocking yourself unconscious, you can't step back, not even for a second.

What you can do, though, is reduce or eliminate certain clear-cut kinds of dopamine stimulation that are problematic, and that's what a dopamine fast really is—a little less dopamine action in certain mental transactions. This can restore your tolerances and expectations back to normal and help you enjoy a more satisfying life. You'd like to feel less frantic, to have more control, and to experience some relief from motivations, attractions, even compulsions, that can be exhausting, overwhelming, and sometimes irresistible. By dramatically reducing stimulation for a time, you give the brain the opportunity to reset what has been damaged by overuse, resensitizing the dopamine system to more subtle forms of stimulation. And by using that time to create new habits and a foundation of knowledge to support better choices, we can make this desensitization less likely to reoccur.

GETTING BACK TO EVEN

The term *dopamine fast* was coined by Cameron Sepah, PhD, a professor of psychology at the University of California, San Francisco.[24] It's a strategy to take advantage of *homeostatic mechanisms the body uses to adjust to changing environmental circumstances*. There's a lot of unfamiliar language in there that's important to understand, so let's unpack it.

The word *homeostasis* comes from Greek roots that mean "to stay the same," which is a high priority for the body. That's why human evolution has created so many mechanisms to keep systems and organs working within optimal ranges. If you get too hot, your body sweats to cool you down. If you get too cold, you shiver, and that heats you up. The eye maintains the optimal intensity of light getting through to the retina by contracting the iris in bright environments and dilating it when the light is dim. These physiologic (bodily) tendencies are what we're talking about when we refer to typical homeostatic mechanisms. Where problems begin is in their frequent overuse.

THE DOPAMINE FAST

For instance, many common foods have sugar added at the factory—bread, pasta sauce, salad dressing, canned soup, even coleslaw—yet we don't find these foods sweet. The homeostatic mechanisms associated with taste have raised our tolerance for sensing when something is sweet. This is another example of downregulation, the same process that occurs in the dopamine circuits and many other biological systems. If you eat a typical US diet, it takes a great deal of sweetness for you to notice sweetness at all. But if you avoided all forms of sugar for even a month, that downregulation would turn to *upregulation*, making you especially sensitive to the thing that you'd been abstaining from. After resetting your homeostatic adaptation this way, a piece of bread would taste more like cake, and many other processed foods would be so sweet you might find them inedible.

Most of the things that give us pleasure are subject to homeostatic downregulation, meaning that when those things are repeated, they become less pleasurable. Lobster was once so plentiful on the Massachusetts coast that it was known as the "cockroach of the sea," a meal fit only for prisoners and livestock. Inmates got sick of it not because of its taste but because they were overexposed to it. Today, though, lobster is a delicious experience in large part because you and I get it so rarely. In the movie *Castaway*, Tom Hanks is stranded on an island where for years he eats nothing but crab. When he is rescued, the lavish buffet at his welcoming gala features what's special to his hosts, crab. He picks it up in disgust, his enjoyment of it having been reduced to nothing.

Which brings us back to the downregulation of dopamine. If you have several windows open and it's suddenly very windy out, you'll close some of them. Fewer open windows means it takes more wind for us to feel its effects inside. Similarly, when the brain experiences a high jolt of dopamine, it "closes some windows"—it reduces the number of dopamine receptors that can catch and respond to dopamine molecules. The more often these spikes of stimulation happen, as with frequent use of certain drugs, the more of these receptors are shut down long term, some forever. Fewer available receptors means it takes more dopamine stimulation for us to feel the kick. Downregulation is the mechanism of desensitization.

It works the other way too. Upregulation is the restoration and repopulation of dopamine receptors. It happens when dopamine-driven stimulation is significantly reduced for an extended time. Unless you've warped the system permanently from extreme use, long-term abuse, or both, the receptor count returns to normal along with our response to stimulation.

This purposeful cutback—not a complete cutoff—is dopamine fasting. The goal isn't to shut down the dopamine system, which is impossible anyway. It's to resensitize the system to more subtle forms of stimulation, which leads to more complex and satisfying experiences of pleasure.

MISTAKES TO AVOID

A 2019 article in the *New York Times,* "How to Feel Nothing Now, in Order to Feel More Later,"[25] is the account of two entrepreneurs who periodically engaged in a day of greatly decreased activity. Their goal was to get as close as possible to the point of feeling nothing, which they believed would hasten the restoration of dopamine sensitivity. "We're addicted to dopamine," one said. "And because we're getting so much of it all the time, we end up just wanting more and more, so activities that used to be pleasurable now aren't. Frequent stimulation of dopamine gets the brain's baseline higher."

On dopamine-fasting days, the entrepreneurs held to strict rules: no eating, no looking at screens, no listening to music, no exercise, no touching another person's body (not even a handshake), no work, no eye contact, and no talking unless "absolutely necessary." The photographer who took their picture for the article wasn't allowed to use a flash because the two were afraid the sudden change in light might stimulate dopamine. Some of this was a good idea, but most of it was useless, even counterproductive.

They were on the right track when they eliminated work. Work is typically future oriented: we invest our time and resources today for income, recognition, or power tomorrow. That's dopamine-driven activity. Avoiding screentime also made sense. Much of what we do on computers and cell phones is planning and abstraction, so screens are a dopamine stimulant, and we can turn them off. The rest of their approach mostly replaced precise choices with blunt force. This took away more than dopamine, and that was costly.

When they deprived themselves of dopaminergic stimulation, they cut themselves off from the pleasures of the H&N system too. As they pushed away the often-empty anticipatory excitement of dopamine, they also pushed away the thing that has to take its place.

To begin with, they cut themselves off from social interaction. Sometimes we interact with others for a future purpose, but more often it's simply for the enjoyment of being with people, a powerful pleasure delivered in part by the H&N neurotransmitter oxytocin. Consider something as simple as a smile. Researchers at Hewlett-Packard used a brain scanner and heart rate monitor to measure "mood-boosting values" for various stimuli. They estimated that a smile was as pleasurable and stimulating as getting 2,000 bars of chocolate or $25,000.[26*] If we smile to help create a better future—to initiate a new relationship, to make ourselves more attractive to a potential employer, or to put a potential client at ease—then the activity is almost entirely dopaminergic. But how often do you present a calculated smile? Most of the time you're just happy. Or consider the stimulation of the exercise they set aside. A runner's high feels so intense that you might assume it's a dopamine-driven experience, but it's not. Runner's high is from endorphin, another H&N neurotransmitter. It creates a feeling of fulfillment and satisfaction, which is the opposite of the urge for more that dopamine delivers.

The combination of all this avoidance denied the entrepreneurs yet another benefit. Dopamine and H&Ns can suppress one another. When you're at dinner, deeply engaged in discussing plans for tomorrow, you're not much engaged with the tastes, smells, and textures of the

---

* Although this study got a lot of attention, it was never peer reviewed or even published, so it should be taken with a huge grain of salt. Still, common sense tells us that smiling likely contributes to our happiness more than we realize.

meal you're eating. Conversely, when you're appreciating the mouthfeel of pasta or the smoky-sweet taste of the cumin in the dish in front of you, debating how to write tomorrow's TPS report is far from top of mind. In this way H&N pleasures can help lower dopamine activity and promote resensitization.

A dopamine fast has to be accompanied by purposeful H&N stimulation, which creates enormous pleasure that doesn't lead to craving. Besides the intrinsic value of in-the-moment enjoyment, and the importance of shifting toward the senses and away from anticipation, the H&N experience reinforces what we need to remember in order to live better: great pleasure is possible somewhere other than the dopamine treadmill. Many wonderful, valuable things are satisfying in the moment and that's enough. By depriving themselves of a wealth of here-and-now pleasures, the entrepreneurs made their "fast" more difficult to maintain, and they missed out on building the habits and experience base they were going to need to permanently lower their attraction to dopamine.

Finally, the entrepreneurs mistakenly tried to modify their brains through an occasional intense day of heroic deprivation. That was never going to work. Some bodily systems can change fast. When it's too dark to see, the iris reacts immediately, but brain systems such as dopamine circuits change slowly. Consider how long it takes to learn a new language or master a new skill. An occasional full day of practice isn't going to do it. Success depends on long-term commitment and frequent repetition. Upregulation is literally the restoration of dopamine receptors at the cellular level. They don't pop up like mushrooms after a rainstorm.

YOUR DOPAMINE FAST

To be successful, your dopamine fast should be built on incremental changes in the short term toward creating permanent habits in the long term. As you plan your dopamine fast, build it around these foundational approaches.

Treat it like working out. Think of this effort the same way you might think of starting out at the gym. Beginners are the most eager to see change. They want to feel healthy and look great, and they want it as soon as possible. Like the fasting entrepreneurs, they adopt the attitude that if doing a little is good, then doing a lot must be better, and exercising to exhaustion is surely best of all. You already know how the story ends: they work too hard on the first day, can barely move the next, then try to forget the whole failed effort. This approach never works. The smart way is to start with small changes and build on them slowly over time. That's as true for changing the way you deal with dopamine as it is for exercise.

Start with small things and do them daily. Reduce dopamine stimulation at predetermined times of the day. Instead of playing a video game or watching TikTok videos, read a novel or talk to someone. Choose some time each day when you avoid sugar or shopping or your Facebook app. Anything that gets you away from the dopamine lure is good; anything that replaces it with H&N activities is better. Just be sure to do the things you've chosen every day. Occasional abstinence from dopaminergic activities won't lead to long-term changes in the brain. Treat it like exercise. It's better to take a brisk walk every day than to do a hard run once a week.

Don't cut back, cut it out. As you choose things to do and to avoid, remember that the most effective approach is to choose a small, habitual thrill that means little to you and eliminate it completely, rather than incrementally cutting back. If you do it halfway, your craving for the dopamine hit will never go away, and eventually your willpower will be exhausted. That's why people who become dependent on alcohol can almost never go back to moderate drinking. The ordinary dopaminergic thrills most people indulge in aren't as addictive as alcohol, but the rule is the same. If you go back and forth, it will always be a struggle.

Eliminating the thrill completely, by contrast, usually ends up being easier than people expect. Do you check your cell phone first thing in the morning? Stop. You might feel anxious and off balance for a little while. The sudden loss of dopamine stimulation might even leave you feeling resentful and deprived. But if you stick it out, those feelings will soon fade. With that victory under your belt, you'll be surprised at how easy the new habit is to maintain. You may even find your former attraction has turned to revulsion. Some people who give up sugar move beyond disinterest to rather strange feelings. They say that after a while, when they look at sugary foods, these former treats no longer register as edible. A piece of cake might as well be a stapler. Their brain tells them that it isn't even food.

Think hard about what's necessary versus what only seems that way. There's no obstacle to completely eliminating drugs and alcohol from a person's life. The same isn't true of some dopaminergic things like social media, shopping, and sex. In these cases, the goal is to distinguish between what is necessary for one's work or personal life, and what's being done solely for the pleasure of anticipating dopamine. Be honest with yourself. As you choose, remind yourself why you're doing this in the first place: to feel better over time. It's a dopaminergic goal well worth some dopamine denial. You'll get better at it every day.

Don't torture yourself with a goal of feeling nothing at all. The entrepreneurs from the article did that. It was a mistake. Feeling nothing is impossible. We're human beings. Don't set yourself up for failure by aiming for an outcome no one could achieve.

Keep in mind that big changes usually begin with unpleasant feelings, and those feelings will fade. The human body resists change—it seeks to maintain homeostasis—so it's going to give you unpleasant feelings or even pain when you try to impose change. For instance, if you stop eating sugar, it can be pretty unpleasant for a while. Along with the resentment and deprivation, cravings might cause depression, anxiety, difficulties with sleep and concentration, and more. But this won't last forever. Some people feel better in a few days, others in a few weeks. And after the body readjusts, something wonderful happens: recovery.

When you walk from a brightly lit room to a dim one, you can't see for a bit, but after a moment or two in the dark you can see just fine. The same is true when we experience other, slower adaptations.

Encourage yourself by remembering the restorative thing that's happening at the cellular level. When you avoid overstimulation of the brain's pleasure circuits, they become more numerous and more sensitive, giving the pleasures you experience greater variety and depth. You get more enjoyment from a handful of berries than you previously did from a glazed donut. You get pleasure not only from the natural sugars in the berries, but also from the multitude of subtle flavors. It's like the difference between the blinding flash of a nightclub strobe light and the blues and pinks that attend the setting sun.

Call it what it is. We need a new name. Instead of *dopamine fasting*, we might call it *dopamine revitalization*. Not a puritanical condemnation of all forms of pleasure, but a way to reject the cudgel of crude stimulation in favor of a nuanced appreciation for the subtleties that make life richer. After years of needing more stimulation to get a good feeling, we're re-regulating to need less. In this way we reorient ourselves to feel *more*. We regain what we have been missing: we make it easier to feel pleasure—easier to be happy. We increase our capacity for gratification and expand the host of things in which we can find pleasure.

RECLAIMING A DEEPER ENJOYMENT OF LIFE

Quick hits of dopamine stimulation from cell phones, Twinkies, and other easy boosts lead to elevated homeostatic tolerance, meaning we experience less and less pleasure from more and more stimulation. To make things worse, that reduced pleasure may increase cravings. The result is an obsessive pursuit that feels, and is, increasingly pointless.

Modern life sets us up for this. Our days are a string of dopaminergic stimulations followed immediately by mostly thoughtless responses. We don't stop very often to reflect on why we're chasing stimulation, nor do we pause enough in the sensory-rich here and now to enjoy what we've earned. That leads to downregulation, and the need for more in order to feel less. But when we pull back from even a few well-chosen dopaminergic stimulations, we break the pattern. Instead of an unbroken string of demands and reactions, each day becomes a string of activities whose value, or lack of value, we more naturally come to recognize. This enhances our power to choose our responses. We can decide to spend less time anticipating things that we likely overvalue, and more time enjoying the undeniable reality of the pleasures of daily life.

We will still experience dissatisfaction. That's not going away. We shouldn't expect it to, nor should we hope it would. Being happy all the time is a problem all its own. And remember, it's not the happy people who improve themselves or the world. But for someone with a revitalized dopamine system, there will be more appreciation from less dopamine stimulation, and more frequent and intense pleasure in the here and now. We feel more human.

Dopamine revitalization lowers our baseline tolerance, making it easier to feel dopamine stimulation. In the end, even a piece of toast can be more satisfying. We begin to experience richer gratification from everyday anticipations: settling in with a good book, looking forward to a healthy meal, catching an unexpected smile from a stranger. The spikes of H&N stimulation become more appreciated delights too: a homemade dessert, a glass of champagne at a party, or recognition for outstanding work. This kind of stimulation doesn't lead us back down the dark path of needing more just to feel less. It restores the color and zest of life.

THE BLESSING OF A CHALLENGE

This won't be easy. Do it anyway.

In Shakespeare's play *The Tempest*, Prospero, the exiled duke of Milan, wants his daughter to marry the prince of Naples. He manages to bring them together and, sure enough, as soon as they meet, it's love at first sight. But Prospero decides to make things difficult for the prince. He doesn't want him to win his daughter too easily since that way he might not value her love. "Too light winning / Make the prize light," he says.

The comedian-philosopher Rodney Douglas Norman recommended something similar in a poem called "I Hope You Have the

Fortune to Experience Misfortune."[27] Rodney says that the best motivation is not difficulty but failure. "I hope you are neglected, ignored, and overlooked," he writes, before wishing us further resistance against the dopamine lure of applause and acceptance: "I hope you learn that you don't need others' approval."

The technological progress that makes our lives ever easier also makes us value lightly things that used to be precious. Rewards that once required hard work and perseverance can now be had by tapping a screen. Social media makes attention easy to find, and online shopping sites bring us by noon what we wanted only that morning. Even calories are cheap and plentiful. In the beginning, the biggest challenge humans faced was staying alive, but that's over. Now gratification comes cheap, and those constant hits of dopamine are dulling our reward systems—dulling our minds, dulling our lives. Now we have to do something that evolution has not prepared us for: manage the ancient circuits in our brain so they don't ruin our lives in pursuit not of survival but of overabundance and constant more. Saying no doesn't come naturally, but in an age of so much plenty, a real challenge can be a gift.

Dopamine fasting may be a recent development, but the larger idea of the power of self-denial has been around for thousands of years. The Greek philosopher Epicurus wrote, "Be moderate in order to taste the joys of life in abundance." He didn't know about the dopamine system, but he understood the importance of maintaining sensitivity at large. It takes some discipline—we have to consciously oppose the evolutionary forces that shaped our brain circuits—but this kind of purposeful living can lead to the restoration of the joy we were made for.

Chapter 4
DOPAMINE TOOLS THAT MIGHT BE COMING SOON

TESTING, TESTING

Once you've worked to raise or lower dopamine and measured your success by assessing your feelings and behavior, you'll probably wonder if there's a way to measure those levels in terms of a number. The problem is that asking about your dopamine "level" oversimplifies what we need to know and what we can even measure—but this may change.

When we talk about medical tests, we're often thinking of a blood test. Blood tests are reliably informative because the bloodstream is where so many important substances are found. But testing blood for dopamine won't tell us much because dopamine levels matter most at the cellular level. For details of your dopamine activity, you'd have to explore the brain with "micron-scale electrodes" and ultrahigh-resolution observation.[28] This has been done in experimental settings and that's about it.

Yet in 2023, a team of five scientists from the disciplines of chemistry, neuropharmacology, and physiology documented their work toward measuring dopamine using a relatively new discovery, carbon quantum dots (CQDs). CQDs are among the tiniest things we know. How small are they? Comparing a CQD to the width of a human hair is like comparing a tennis ball to the length of two football fields. It turns out that when things are that small, the rules of physics get weird. Because of that, CQDs can be made to light up, to deliver medicines to certain brain structures, and to react to certain molecules. They're also fairly harmless in the body, so they're ideal for brain research and the measurement of function—they are stable and "keep to themselves," meaning they are largely resistant to whatever is around them. That means we can introduce them into the body without creating a problem—and since testing requires getting close to the thing you're measuring, CQDs are a potentially powerful tool. In a carefully prepared environment, the radiation emitted by CQDs is proportional to the amount of dopamine around them. The important phrase here is "carefully prepared environment." At this time, it would be nearly impossible to create such a setting in a medical office or even a hospital. Still, this could be a first step toward a precise test, enabling doctors to diagnose dopamine-based disorders at the molecular level.[29]

Another test of dopamine levels is already out there, but whether it works remains an open question. It is breathtakingly simple, and when I say simple, I do mean simple: it's a dopamine scale based on a person's eye-blink rate (EBR), calculated by nothing more than counting the number of spontaneous eye blinks per minute. A 2017 survey of the scientific literature on the method's accuracy described it, at that point in time, as "inconsistent and incomplete."[30] Yet there does seem to be something to it, at least sometimes and for some people. Plus, the EBR test has helped in the development of another potential dopamine enhancer: a very unusual sound.

DOPAMINE TOOLS THAT MIGHT BE COMING SOON

LISTEN TO THIS

Can we alter the effects of dopamine by slipping on a pair of headphones? In a 2013 paper, three scientists documented how they used EBR to measure dopamine levels while creating an "auditory illusion" known as binaural beats.[31] The effects were encouraging.

To understand how we might change how we feel through sounds we hear, we first need to understand a bit about waves. Many things in nature carry wavelike qualities, including the repeated patterns of electrical activity in the neural tissue that makes up the brain. But to understand how waves affect the brain, let's consider something we're already familiar with—waves in the ocean. There are many kinds of ocean waves, from rolling waves to gentle ripples. Rolling waves are large and slow. They carry a lot of energy and, over a short period, significantly affect the shape of a shoreline. Gentle ripples are small and fast. Their effect on the shore is slower and more subtle. Like ocean waves, brainwaves are associated with specific actions and outcomes.

Think of the frequency of waves as a pulse, like the beat of a song. For example, the Bee Gees' "Stayin' Alive" is about 104 beats per minute. Brainwaves have frequency, too, and each range of brainwave frequencies is associated with an activity or state of mind:

- **Delta waves** occur from once every 10 seconds up to 3.5 times every second (0.1–3.5 hertz, or Hz). These waves are associated with dreamless sleep and infancy.
- **Theta waves** (3.5–7.5 Hz) are associated with creativity, fantasy, and intuition, as well as a lack of focus.
- **Alpha waves** (8–12 Hz) are the most common state for the brain. They occur during simple, casual awareness. If you're awake and not busy, you're probably producing alpha waves.
- **Beta waves** (12–30 Hz) are the next range up the scale. They occur when you're occupied, making decisions, paying close attention, or thinking hard.

- **Gamma waves** (30–44 Hz), the highest frequency commonly observed, are associated with simultaneous processing across multiple areas of the brain.

Now consider sound, which is a wave shape through a medium like air, repeated over and over. A gas engine in a car makes a sound at a somewhat low pitch, but when you rev it, that sound has a higher pitch. Why? The engine is making more RPMs, or revolutions per minute, and that higher frequency of activity creates a higher-pitched sound.

With all this in mind, we can examine a powerful oddity of brain function. If we play a pitch in one of your ears and a slightly lower pitch in the other, you will hear a pulse or "beat" with a frequency corresponding to the difference between the two notes. For instance, if one is the musical note C at 132 Hz, and the other is A at 78 Hz, you will hear* a beat that occurs at the difference between them, 54 Hz.[32] This brings us to a surprising outcome: *Various frequencies of binaural beats appear to change the frequency of brainwaves and associated brain function, impacting mood, feeling, and cognition.* All we have to do is precisely choose a pair of tones we listen to, one for each ear. The difference in frequency pushes our brainwaves to that frequency and stimulates related activities in various structures and systems.

One of the things we can stimulate this way is dopamine. The scientists behind the 2013 paper studied the effect of binaural beats by examining creativity. The result: Binaural beats seem to improve the brain's ability to gather many different ideas in a relatively short time,[33] a skill called "divergent thinking" that we'll return to in chapter ten. This collecting of disparate elements, a particular kind of brainstorming, is the first step in purposeful creative thinking, because new ideas come from the connecting of otherwise unrelated things. In terms of the influence on dopamine, the results suggest that individuals with low dopamine benefit most from binaural beats and

---

* Sometimes the beat is too low frequency to consciously hear. The brain still senses it, though, and reacts.

those with normal or high levels do not. Although, in another example of the brain resisting "overclocking," members of the normal- and high-dopamine groups may hear binaural beats and experience decreased creative capacity.

How this works is unknown, but there's something interesting going on. Binaural beats may someday help us be more creative when we have a deficiency of dopamine activity in that related part of the brain. There may even be a future in using auditory stimulation to affect dopamine levels to affect mood, feeling, and cognition. It's likely a long way away, but the results so far are intriguing.

IN SEARCH OF THE EASIER WORKOUT

We would all like to feel more motivated. Here's how we may someday achieve this via dopamine.

Manipulating microbiomes in mice can boost dopamine activity and motivate them to exercise. We also know that human beings with relatively high dopamine levels are generally more willing to exercise—but that's true only when there's a reward in front of them.

Is it possible for dopamine to motivate us—in this case, to exercise—without the offer of a reward?

Researchers began with nineteen volunteers already diagnosed with low dopamine because of Parkinson's disease and followed them through three experiments. Some days the subjects took their usual dopamine replacement. Other days they were told to skip the medicine. Over several days, researchers directed the subjects to squeeze a hand grip that measured the pressure exerted.

Experiment 1

Researchers asked the test subjects to squeeze the hand grip and rate the difficulty.

Results: When the subjects took their medicine to normalize

dopamine levels, they accurately reported the resistance of the device. When they didn't take the medicine and had low dopamine, they overstated the resistance.

Conclusion: *When dopamine levels are below normal, a physical task might be anticipated as more difficult than it really is. That could discourage you from attempting it.*

Experiment 2

In the second experiment, researchers had some patients take their dopamine-restoring medicine and had others skip it. They then gave the subjects a choice:

- Option 1: Squeeze a grip that is guaranteed to be easy.
- Option 2: Flip a coin. Heads, squeeze a grip that requires no effort at all. Tails, squeeze a grip that will require more effort than usual.

Results: Subjects with normal dopamine levels were more likely to take a chance on the coin flip and risk the tougher task. Those with low dopamine were more likely to take the easy option.

Conclusion: *When dopamine levels are below normal, you may have less confidence in your ability to complete a physical task, so you are less likely to attempt it.*

Experiment 3

In the third experiment, researchers gave participants a choice:

- Option 1: Accept a small amount of money and be relieved of participating in a physical task.
- Option 2: Flip a coin. Heads, you get no money. Tails, you get more than the Option 1 amount.

DOPAMINE TOOLS THAT MIGHT BE COMING SOON

Results: There was no difference in the responses between the medicated and nonmedicated groups.

Conclusion: *The influence of dopamine on increasing confidence appears to be specific to physical activity.*[34] *It does not seem to have an effect on generalized confidence or nonphysical risk.*

This and similar experiments suggest that someday there may be a dopamine-related treatment that could make us perceive a workout (or some other demanding task) to be less exhausting than it is, and thus make us more inclined to exercise. That would be a valuable bit of motivation for a world in which obesity and the sedentary lifestyle are so common and costly.*

CREATIVITY VIA THE WALL SOCKET

Technology gives us the ability to do all sorts of things faster and easier than we used to. Calculators and computers enable us to perform complex calculations that would take much longer by hand. AI can produce a time-saving summary of anything from an email thread to an entire book. Might there be a technology to boost our creativity too?

Recent work suggests that such an outcome may be possible, even imminent. It also may spare us the downsides that typically come with treatments that "overclock" an otherwise normal brain. There are already several ways to stimulate the brain into improved performance by using electricity. The most well-known is electroconvulsive therapy (ECT) or electric shock therapy, used for treatment-resistant depression. Another is transcranial magnetic stimulation (TMS), which is similar to ECT except, instead of routing electric current directly through the brain, it uses a magnetic field to create more subdued currents. But the brain stimulation technology we're interested in is so new that it has been investigated only minimally.

---

* There's still no research toward making a workout fun, though my gym has free bottled water and really big TVs, so there's that.

Transcranial direct current stimulation (tDCS) is a more precise tool than ECT or TMS, and it seems to do via electrical stimulation what we can now do only via medication. For instance, Xanax has a sedative effect by stimulating the release of GABA, a neurotransmitter that blocks certain stress-making signals in your brain. tDCS does the same thing but with electricity* instead of neurochemistry, though the effect isn't as strong as medications. The method is noninvasive. By carefully positioning electrodes on our head, we can choose the structures we want to stimulate.

What if we could boost creativity by using tDCS to inhibit the control dopamine circuit? That would be useful, since the circuit blocks odd or seemingly unreasonable ideas, which are often useful when we're pursuing new approaches and fresh designs. The control circuit involves the prefrontal cortex, which is right behind the forehead, making it the only element of the control system we can reach with tDCS stimulation. Scientists placed a negative electrode over this area, hoping to inhibit the control circuit and make it easier for odd ideas to get through to the desire circuit, home of creative thinking. In an experiment, volunteers stimulated in this way were able to come up with more alternative uses for everyday objects than those who were not stimulated. In fact, in the estimation of the experimenters, the subjects' ideas were especially odd or uncommon.[35] The experimenters equated this with a boost in creativity. Normal ideas still came through, too, because they weren't being blocked in the first place. This matters because we don't want to shift our creative efforts to only the production of nutty ideas. (We'd like only a sprinkling of such nuts.)

At this time, tDCS isn't being used on patients. It's just a research tool to manipulate the brain in interesting ways to see what happens. However, it uses a fairly simple technology, and at the time of publication you could buy a tDCS device on Amazon. Some graduating

---

* A receptor has a positive or negative charge relative to the fluid it sits in. Introducing electricity into this environment changes that charge, and the charge is what determines whether a neuron fires. It's all very interesting if you're into that sort of thing, but you don't have to understand these details to get the point I'm making here.

DOPAMINE TOOLS THAT MIGHT BE COMING SOON

students presented one as a gift to my friend and partner on *The Molecule of More*, Dan Lieberman. He has it on display in his office but he hasn't been brave (or foolish) enough yet to strap it on. I have asked him to let me know when he does, as it could make for a fun afternoon.

It's reasonable to believe that a version of this technology for boosting creativity is in our future. It's probably just a matter of time, testing, and a little more research—plus finding someone to test it.

WHAT IF IT COULD GIVE YOU BACK YOUR LIFE?

Many dopamine-related treatments now in use deliver their benefits at the expense of our ability to enjoy some of the simplest pleasures of life. For instance, many antidepressants have unwanted sexual side effects.

What if we could fix that?

There's a compound under study right now that might get us there, though the details sound less like medical research and more like the origin story from a superhero movie.

Here's a fact not widely discussed outside of pharmacological research: many drugs start out as poison. Penicillin, as you've likely heard, comes from mold—something you don't want to ingest. Another example: Nearly a hundred years ago, veterinarians noticed that injured cattle chose to eat a certain clover. Usually, eating the clover killed them. When they were sick, it helped them. An extract from the plant proved to have a medicinal use in humans and became the blood thinner warfarin.[36] Lithium, one of the first breakthrough psychiatric medications of the twentieth century, was developed by Australian psychiatrist John Cade. He figured out its power by working with an extensive collection of his patients' urine.[37] (He kept the jars in his refrigerator. The next time you think your fridge is toxic, think of Dr. Cade's kitchen.)

The latest example of this age-old story features beta-carbolines, implicated in Parkinson's disease. In high concentrations they can attack the dopamine system by tearing it apart, neuron by neuron. But if we slightly modify a beta-carboline,* something surprising happens: it switches from destroying dopamine neurons to protecting them, creating what two Russian neuroscientists characterized as "good guys from a shady family."[38] The rehabilitated neurons improve oxygen use by the targeted cells, ultimately promoting the creation of more dopamine. The restored cells also survive better under stressful conditions, proliferate more easily, and make better connections to other cells. The end result is improved function for the patient.

As we've seen, fooling around with the dopamine system can have bad effects too. Lab tests with these modified beta-carbolines in animals have shown improved learning and faster adjustment to new environments, but that's no guarantee for what might happen with humans, and there don't seem to have been any human studies so far. But that doesn't mean people aren't trying it anyway. Since this approach is not yet regulated by the FDA, the general public can do with it what they like,† and some of them have. The results are intriguing. Here's how one user described it:

[This molecule] was what I used after I felt years of ADHD medication had made it difficult to feel natural world stimuli, and it worked within a few days. I was out enjoying sports again and basic socialization. [It's a] subjective perspective but a lot of these anecdotes [are] online in that community.

The effects maximize within the first week and stabilize thereafter. It's not a "high" and only feels mildly stimulating. But you begin to enjoy simple things like walking outdoors in the sun all over again or seeing a colourful plant. The effect did not wear off; it just became my new normal so to speak. Even years later.[39]

In some ways the current evidence for this kind of treatment is similar to that available for nutritional supplements, whose claims rely

---

* We shift a carbon atom from one position to another in the molecule; again, you don't need to know this to understand what's going on but I include it for completeness.

† You'll notice I have not identified the compound. It's legal to experiment with it outside the lab but the risks are profound.

on animal studies, anecdotal reports, and not much more. But setting aside the need for a greater sample size, the animal studies are credible, starting with the fact that they were performed by individuals with no financial interest in the outcome. As for users' comments, they're looking for help, some of them desperately. If the compound didn't help, they had no reason to say otherwise. In fact, they would have had every reason to talk down the drug, having been let down yet again.

Does that mean people who want to reinvigorate their dopamine circuits should take it? We don't actually know that it works, or if it's safe. We've seen cognitive improvements at low doses in animals, but at higher doses it has triggered anxiety behaviors and seizures. People who didn't harm themselves permanently by taking it may have just been lucky, and those who really did feel better may have just gotten *very* lucky—or maybe they're the first users of what will turn out to be a miracle drug. Either way, who it can help, how much it can help, and why, if it does at all, is a roll of the dice just now, and dosage is a guess. Except for the terribly desperate, the risks far outweigh the possible benefits. But inside the lab, and in well-controlled human trials, this is a promising path of study, and scientists want nothing more than to make it work so people can get some relief. Stand by.

THE REPLACEMENTS

There's one thing that would be better than introducing outside-sourced dopamine to a brain that doesn't produce enough: restoring the brain's ability to create it. We could, in theory, transplant precursor neurons into the brain, which would develop into complete cells, start making dopamine, and then keep doing it as if they'd been there all along. These kinds of implants have been the distant dream of doctors for nearly three decades, conceived after the first isolation of embryonic stem cells that could, in principle, do the job. There's been little progress because the work is wildly complicated, it's bound up in ethical questions and international legal coordination, and it's risky to the test subjects. But a group of researchers led by Claire Henchcliffe, a neurologist at the University of California, Irvine, seems to have taken the first step toward a practical treatment. If progress continues, it could lead to the most powerful curative in the history of brain science and the restoration of normal lives for millions with brain injuries and diseases, including Parkinson's.

The front-and-center issue is safety. When implanting these cells in humans goes wrong, it can go very wrong. One test in the year 2000 was "nightmarish": the newly implanted cells went wild and could not be turned off. Younger Parkinson's patients experienced horrifying symptoms such as constant chewing, flailing fingers, and arms flinging about. "It was tragic, catastrophic," said Dr. Paul E. Green, one of the researchers in the study.[40]

Twenty years later, Dr. Henchcliffe's work directly addressed this problem. After animal studies showed excellent safety and tolerability, Henchcliffe's team received permission to proceed to a first-in-human study. In a small test of twelve people with Parkinson's disease, the team implanted lab-made neurons that produce dopamine, whose shortage is key to the condition. Not only did the new cells produce replacement dopamine, they seem to have become a working part of the brain, staying active for about eighteen months so far, as of March 2024.[41] These implants, a treatment known as MSK-DA01, "appear to be safe and may have reduced symptoms" for some participants.[42] This is the promise of regenerative medicine: instead of patching the problem, working around it, or creating a substitute, the brain literally acquires a fresh version of the structure that failed. It's the difference between, for example, attaching a prosthetic limb and growing a new one. The CEO of the company behind the effort said, "There is a day when we hope that people don't think of themselves as Parkinson's patients."[43]

The Henchcliffe study was a phase I trial, which means it was not about efficacy but about safety, side effects, and dosing. A phase II trial will produce hard data on whether the treatment really works. If that effort is promising, a larger phase III trial will compare the outcomes to those of current treatments. If the new treatment appears to be superior, the last step will be applying to the FDA for approval. This work is very preliminary and there's a lot of effort ahead, but it's promising. But if the treatment works, it could be a path to supplementing or replacing neurons in other dopamine systems in the brain.

THE END OF ADDICTION?

Some conditions we can cure by eliminating the problem wholesale. Malaria responds to artemisinin* drugs. Cholera is solved with antibiotics and rehydration. A couple of aspirin often fix a headache. Other conditions can't be made to disappear, but we can end their harmful effects permanently with simple, ongoing treatment, such as diuretics for high blood pressure or hormone therapy for hypothyroidism.

What if dopamine-driven cravings and urges could be made to go away too? What if we could end addiction?

In modern medicine no one is cured of addiction. People acquire ways to diminish the worst of the urge, then learn to manage what remains. That's an excellent outcome but it's not the same as a cure. The problem is bound up with emotion, mood, and craving, which are just as often part of pleasant things. To "cure" addiction we'd need a compound that not only targeted particular aspects of the human experience but also had an effect only in certain situations. In particular, we'd need to tame cravings, which dopamine helps cause. With overindulgence over time, pleasure fades but craving does not—that's the power of addiction. But on that fact hangs a potential solution.

Semaglutide is the key ingredient in brand-name medications including Ozempic, Rybelsus, and Wegovy. The original use of semaglutide was to treat type 2 diabetes, which comprises about 90% of diabetes cases. Semaglutide helps control blood sugar levels. That's what counts in diabetes treatment, but it does a couple other things too. It slows the movement of food through the stomach and decreases the desire for food. The result of this has been not only better diabetes

---

* What's artemisinin, you ask? Why, it's no more than a sesquiterpene lactone molecule containing an unusual peroxide bridge. Or in plain language, it's a Chinese herb that seems to have antiparasitic properties.

management but also weight loss. This side effect appeared in pretty much anybody who took it, diabetic or not. They didn't even have to try that hard to cut back on what they ate. They naturally ate less. That's the key to the success of semaglutide: it makes the craving for food decline to nearly nothing. Already, some businesses are adjusting their plans to account for how semaglutide may change culture and commerce—for instance, Weight Watchers is shutting many physical locations, apparently in part because of this new approach to diet and lifestyle. That's a big deal. But even bigger is that these new weight-loss drugs may have historic implications for addiction at large.[44] This experience of "the end of craving" can be profound, and I have not had to look far for an example.

I've battled obesity most of my adult life. I've lost weight but always with great struggle. My problem is simple and if not a true addiction then reminiscent of it: I crave the experience of eating. I can be full and still enjoy it. If someone asked me if I wanted to order a pizza, my answer was always *yes*. I don't recall considering if I was hungry. I like the way it feels to eat. I like the flavor of food. I like the experience of eating. I have declined weight loss surgery because it limits forever the ability to indulge in this thing I love. Not be able to tuck into an extravagant meal anymore? Ask if I'd like to give up breathing.

When I became a semaglutide patient, the drug diminished not just my appetite, a result which I anticipated, but also my craving for food, which I did not. I was told this would happen; I just couldn't imagine such a thing could be real. But it was—and it was amazing. Then came an even bigger surprise: it also took away my desire for the pleasure of eating—the infinitely satisfying H&N act of consuming something delicious. Imagine forgetting someone you were in love with. It happens. But imagine not caring if you ever fall in love again. This, unbelievably, was like that.

I gave myself my first injection on a Monday at 9 AM. Three hours later I attended a business lunch at an Italian restaurant that I love—a new one, so there was still a lot of dopaminergic anticipation attached to it for me. I'd barely worked through the menu! Yet not only did I lack an appetite, the food didn't appeal to me at all. Just to have something in front of me while taking up a table, I ordered a side dish and picked at it. It's been like that ever since: *Food? Nah. Not much interested.*

The drug makes me feel full all the time, but feeling full had never before stopped me from eating. The difference has been that I no longer crave food. Ask me now if I want a pizza and I'll tell you *no, I don't.* The desire is just . . . gone. The intense pleasure of eating? That's gone too. Things still taste good, but the absolute urge that will not be denied—that drag-you-by-the-neck demand to eat? It's evaporated.

Your rapidly slimming correspondent is not alone in this experience, and for many, this decrease in desire extends to more than just food. There are reports across Great Britain, for instance, of semaglutide users experiencing a decline in far more insidious cravings, such as those that come from the chemical addiction to cigarettes and alcohol, and behaviors such as nail biting, skin picking, and compulsive shopping.[45] Here's why that may be happening.

The connection between dopamine and craving begins at the pancreas.* Semaglutide causes the pancreas to release insulin over a long period. Similar agents such as liraglutide and exenatide act over shorter periods. All of them do this by stimulating a receptor for the hormone GLP-1, which stands for glucagon-like peptide-1. A compound that stimulates a receptor is called an agonist, hence these compounds are "GLP-1 agonists." In addition to stimulating the release of insulin, these agonists ultimately cause action along the dopamine-reward pathway. This is where the promising effects begin: studies from 2018 to 2020 on rats and mice showed GLP-1 drugs had a mediating effect on dopamine stimulation from alcohol,[46] cocaine,[47] and oxycodone.[48] Another study found semaglutide "promoted the growth of dopaminergic neurons in the substantia nigra" in mice.[49]

It's a long way from laboratory studies in animals to safe and effective treatments in humans, but the search is underway. Much of semaglutide's method of action is unclear. Perhaps it ultimately creates some inhibition in a precise part of the dopamine system, the part we associate with cravings. That's a very general description, like saying,

---

* This explains why the potential wonder drug grew out of diabetes research.

"Perhaps we can sail to Europe because this boat tends to float," but it's a place to start. There are many moving parts to explore. Semaglutide may work by increasing the number of dopamine receptors, which could diminish compulsion or addiction because such a system is more easily satisfied. Then again, more receptors could instead drive increased compulsive behavior since the pleasure from stimulation may be greater. We also know that otherwise similar receptors and neurons in any such system can yield wildly different effects depending on where they are located. For example, the chocolate-chip-sized nucleus accumbens is where dopamine is released in the desire circuit. When dopamine is released in the shell of the nucleus accumbens, the feeling is reward; when dopamine is released in its core, the feeling is craving.

We know only this for sure: if semaglutide and other GLP-1 agonists' very promising effects on addictive behavior can be associated with particular actions at the cellular level, it may lead to an effective and profound treatment for chemical and behavioral addictions and compulsive behavior. That would be a dopamine effect to transform the world.

Chapter 5

RAISING AND LOWERING DOPAMINE THROUGH THERAPY

TALK, LISTEN, CHANGE, IMPROVE

If you want to change the overall effects of dopamine in your life, yet another path is possible: therapy. Consider focusing less on understanding the neuroscience of dopamine and more on working with your feelings and behaviors—a more traditional way to build the mindset you seek and a life closer to what you'd like it to be. Therapy is powerful and proven. Whether you're dealing with too much dopamine activity, not enough, or, more likely, issues that are more subtly involved with dopamine and its interplay with other systems, therapy relieves you of having to understand brain chemistry and allows you to focus directly on mind, mood, attitude, behavior, and choice.

For those willing to put in significant effort—to treat it like a job with value in both the work along the way and the goal at the end—therapy can be a peerless exercise to improve life. But you have to commit to it the way you do to your family or a career you love. Therapy can make your life better on all fronts: your approach to problems, the way you accommodate disappointment, how you manage urges, how you treat other people, and the way you make decisions. Engaged therapeutic intervention can be the path to finding satisfaction and meaning in life. It's that useful.

For people who are especially self-disciplined, some therapeutic methods can work as a self-directed effort, no therapist needed. This underlines the fact that therapy is more than a conversation. Within a framework of guidance derived from science and philosophy, it is engagement with yourself toward a frank consideration of who you are and who you want to be.

Talking is a part of therapy but it's not the only part. There has to be more. Simply talking about a serious problem with behavior, mood, and outlook is about as impactful as talking about a broken arm. The work must go deeper, such as defining what qualifies as a problem; identifying what is and is not a problem for you in particular; and reckoning, sometimes in uncomfortable ways, with why you have certain feelings and engage in certain behaviors. Therapy requires training oneself to accommodate discomfort. This lets us better estimate how much of what we feel is a response to real threat and how much is only ingrained reaction to stimulation. Finally, therapy requires us to take action toward improvement, which can also be uncomfortable and even frightening. Therapy is a multifront effort to improve how we feel, behave, react, and conduct ourselves. It must have a goal, it must be built on a method or theory, and it must result in regular progress. Therapy is improvement made manifest by action.

The most commonly used therapeutic method just now is cognitive behavioral therapy (CBT). It is the most researched and studied form of psychotherapy, its outcomes tested in quantitative terms against objective standards. Most important, its results are typically more positive than those of other approaches.[50] CBT aligns with current consensus on both how the brain works physically (neuroscience) and why we feel and behave the way we do (psychology). It's not perfect and it's not complete, but there are a lot of people living more successful, more satisfying lives today because they pursued CBT and stuck with it. Practitioners and experts will argue about details of its method, but what matters to you is whether it can help, and it absolutely can. CBT, often in combination with medication, can provide relief from many common, troubling psychiatric issues, including dopamine-related problems such as depression, panic, anxiety, phobia, bipolar disorder, bulimia, obsessive-compulsive disorder, and posttraumatic stress disorder, or PTSD.

You can pursue CBT on your own, but a self-directed approach is not for every person or every condition. If you're severely depressed to the point that you don't go out and you're missing work, trying CBT by yourself is probably too heavy a lift. But if you have mild to moderate symptoms of depression or anxiety, or you can clearly identify the behaviors and feelings you want to change (and they're not profoundly preventing you from living life), you can likely find a lot of help in a CBT-based book or online program.[51]

To practice CBT, you're going to undertake a methodical examination of how you think and feel. From the understanding you gain, you'll propose connections between the things you believe and why you believe them, which will help you decide whether what you believe is worth holding on to. You'll practice identifying and classifying your thoughts. You'll seek (and find) connections between those thoughts and the feelings that accompany them. One of the most helpful things you'll do is test those thoughts against reality, then practice replacing false ideas with true ones.

Here's a way to do that, and an easy way to remember: call it "possibility versus probability." For example, a common troubling thought is that someone who is late for an appointment with you has been in an accident. While that's possible, it's more likely they have been held up for a benign reason. Consider how often accidents happen versus harmless delays. What's more likely?

Another example: You feel that your lack of success at something is because you have bad luck. In fact, you can probably examine your behavior and circumstances to identify concrete causes for your lack of success as well as ways to improve the situation.

For those seeking a more personalized approach to taming the challenge of dopamine, engaging with a reputable therapist is a great choice. In the wake of the pandemic and the rise of remote meetings with medical practitioners, more therapists are available than ever, often at a lower price than they used to be. An engaged professional can answer your questions and guide you through your self-inquiry. They'll also act as an accountability partner, further guaranteeing that you'll seriously engage with the hard work that success at CBT requires.

While understanding the role of dopamine in the challenges you face will help you get more out of all this, that's not because therapy depends on your understanding of neuroscience. It's because that knowledge will have given you a head start on serious thought about your urges and motivations. Dopamine is at the heart of desire, urge, reward, and pleasure—and CBT provides a path to gaining better control over them.

ACCEPTANCE AND COMMITMENT

"There's no such thing as bad weather, only the wrong clothing."[52]
Billy Connolly, comedian

CBT is built on the idea that if you understand the reasons for your thinking and behavior, you'll be better able to change them. Traditional behavior therapy is built on the idea that all behaviors are learned, so we can change them by learning something else without examining our underlying motivations. Both have a sound record of success, though CBT is far more widely used.[53] But another approach is gaining ground, one that, as Billy Connolly observes, is more about changing our reaction than our situation.

In the 1980s, Steven C. Hayes, a professor of psychology at the University of Nevada, Reno, proposed what he called Acceptance and Commitment Therapy (ACT, pronounced as the word *act*). ACT incorporates elements of CBT and traditional behavior therapy. However, instead of identifying troubling thoughts and feelings, then helping patients understand their source and acquire the tools to eliminate them, ACT directs us to accept that certain kinds of troubling feelings are a part of a healthy emotional landscape, and that in many situations our discomfort with them is natural. Therefore we will be better off learning to accept them, the better to pursue our lives.[54]

This is an especially valuable approach for feelings that arise from dopamine. Why? We could not function without dopamine-driven motivation, even though the price of it often includes difficulty and pain. Attempts to rid ourselves of dopaminergic motivation will be doomed to fail, because dopaminergic motivation is key to many things that aren't problematic at all and are necessary to life. ACT reorients our minds around the idea that not every discomfort is best eliminated. Some discomforts, including many that are dopamine driven, are a part of the yin and yang of life. ACT gives us a way to act on this truth and make progress.

Let's say I buy a house with a huge pole in the middle of the basement. This pole prevents me from using the space for a game room. I have several choices:

- I figure out why the pole is there. Maybe it's a supporting pole that holds up the house. Maybe it used to be a supporting pole but isn't needed anymore. Maybe it's decorative. Or maybe it's something I have yet to think of. Whatever it is, I'll have to work with an engineer or at least a carpenter to figure it out. It'll take time and resources to do so, but with this knowledge I may find a way to get rid of the pole, though I might not.
- I get rid of the pole without investigating why it's there. This may work out or it may not. I may get the game room I wanted, no problem, or I may get it for a while only to end up with related problems later, like the living room collapsing onto my new pool table.
- I accept that the pole is a part of the room. I see that it was part of the necessary structure of the house. I look for ways to work around it. This has the advantage of allowing me to begin using the space sooner rather than later, although I may not get to use it in the way I originally planned.

The first option is like CBT: let's take the time to figure out why the problem exists, then act on that knowledge. The second is like traditional behavioral therapy: I don't want to spend time and effort figuring out why it's there. It's possible we won't figure it out anyway, so let's just get rid of the barrier and move on—and accept that we may encounter unintended consequences. The third is ACT: the pole is a part of an immutable landscape. Let's choose to accept that and live accordingly. If we do that well, in time we may no longer notice the problem. We may not even see it as a problem.

ACT has unique and practical qualities that make it attractive to those taking on dopamine-driven problems. What recommends it most is its direct focus on helping people feel better right away: ACT turns our attention immediately to improving our state of mind by modifying how we behave. There's no waiting while we engage in deep self-examination to identify some troubling experience and untangle it from our behavior. CBT requires us to identify and reject ideas that influence our lives. While that's a powerful approach, it can take a lot of time. ACT builds upon our existing mindset to directly move us toward our desired outcome.

ACT isn't for everybody. It requires learning to live with discomfort that you may believe is the fault of others. Some people can't do that. They may be better off with CBT or behavioral therapy, both of which take a problem–solution approach that begins at the source of the trouble. ACT starts from the perspective that life is an experience of many kinds of feelings, not all of which are pleasant, and that accepting this variety is key to finding greater satisfaction and meaning, and more day-to-day happiness.

Why do we prioritize digging into problems, sometimes seeking to place blame over finding relief? Here's one theory. Technology creates progress by asking us to identify a problem, then to derive a solution. This approach is now so ubiquitous that it has been stretched beyond where it works into places where it may not, such as the mind. Many of us now think of every discomfort as a problem in need of a solution—a new idea that draws us toward perpetual conflict with others and ourselves.

"[W]ith every scientific and technological advance, a discrepancy-based mode of mind grows stronger, and our ability to be present, aware, and flexible grows weaker," Dr. Hayes wrote in 2009. "Yet we as a culture seem to be dedicated to the idea that 'negative' human emotions need to be fixed, managed, or changed—not experienced as part of a whole life. We are treating our lives as problems to be solved, as if we can sort through our experiences for the ones we like and throw out the rest."[55] ACT promises a faster way through the challenge to reach the richer life that results. While it requires us to accommodate difficulties, the benefit of this is typically a more fulfilling experience of emotions and perspectives, greater empathy, and a deeper tolerance for our own shortcomings and those we see in others. ACT steers us toward finding meaning in all experience, and away from elevating pleasure and comfort above all. It shows us that a satisfying life is built on more than delight, just as a healthy diet is built on more than what is sweet.

ACT: SIX SKILLS

ACT is built on six skills that promote psychological flexibility in experiencing, processing, planning, and reacting. As with physical flexibility, we build these capacities through practice and (mental) exercise. Some steps promote fluency in how we deal with anticipation. The others promote our appreciation of the here and now. Taken together, they build harmony between our accommodation of what we feel today and our desire for what we want in the future. They can guide us to achieve the balance we need to constructively address dopamine-driven problems.

Skill One is Acceptance. When I was an undergraduate student in physics, the first thing my instructors taught me was to inventory all the facts at hand without regard to my opinion of them or their apparent utility. That's what we do in this step: we strive to become aware of what we are feeling, the better to know ourselves fully and without bias. We acknowledge difficult feelings as well as pleasure. We recognize when we're denying feelings, masking them, or avoiding them. In this way we begin to teach ourselves to experience the world as it is so we can build the internal life we want.

Skill Two is Cognitive Defusion. You need to be able to separate or defuse feelings from your reaction to them. This demands self-awareness in uncomfortable, even painful moments, and it does not come naturally. Yet learning how to do this will give you the ability to reject being consumed by fear in favor of consciously and purposefully choosing how you will respond.

Let's say you're driving down the road when a terrible storm sets in. Your first reaction is pure feeling: *It's hard to drive. The road is slick, and I can't see very well. I'm getting nervous, and that's making this even more difficult.* But in a moment another thought may occur to you, a thought that is not feeling but considered reflection: *This is a rainstorm. Rain makes the road slick, so my reflexes need to be as good as the situation requires.* Based on this assessment, you may decide to slow down, pull over, or continue to drive, just with more caution.

Intense feeling is a natural response to stress but it doesn't have to be the basis for reaction. Instead of fusing the experience of the rain with our reaction to it, we can separate or defuse them. The rain is *stimulus*. What we do in response is our *reaction*. Let's put a gap between them: we experience the feeling, then pause to remember that we can reject immediate, unconsidered reaction. Now we've improved things for ourselves. First, we've given ourselves time to practice cooler, more useful observation. Then we've better equipped ourselves to come up with a thoughtful response. Cognitive defusion creates a space for a back and forth between here-and-now processing at the time of stimulation, and the dopamine-driven capacity for calculation.

Skill Three is Performing as the Observing Self. This is what's going on in the time between experiencing some stimulation (*The highway is slick!*) and taking some considered action because of it (*Slow down!*). That purposeful response is made by the *observing self*.

It may be the result of simple reasoning, such as deciding to slow down on that wet highway. That's an improvement over thoughtless response. But in the face of some challenges, the process may be deeper and more personal. In those cases, we process emotions and facts, create possible courses of action, then, by way of mental time travel, evaluate the possible results. We then consider all that in light of the self-image to which we aspire. Finally, we make a choice that comports with who we are and who we wish to be.

Let's say I'm talking with someone about politics, and they've shifted from matters of policy to a barrage of personal insults. Let's further stipulate that I've become pretty good lately at defusing stimulus from reaction, and letting the observing self do its work in the time between—in a moment of stress, I have the ability to distance myself from my emotions and choose a more considered reaction. I factor in the facts, my true feelings, and how various reactions comport with whom I consider myself to be.

If my "list" of self-beliefs is *I'm an open-minded person, I'm pretty patient,* and *I like to learn new things*, I may ignore the insults, challenge my friend's way of arguing, or ask him to stop. I might also decide that there are exceptions to when I act on these self-beliefs and instead just walk away. Or I might be led to an entirely different reaction. How tightly or loosely we hold to our self-beliefs is a measure of how easily we can change—for better or for worse.

Whatever I choose, the work of the observing self is almost always superior to simply reacting. We make better choices when we favor our analysis of emotion over the emotion itself. Call it the satisfaction of self-control. The observing self lets us weigh desires about the future against what feels pressing in the present. It's an exercise in productively using the best of our dopamine-related qualities in concert with how we feel in the here and now.

Skill Four is Living in the Present. Not to be confused with recognizing feelings as feelings and processing them from some mental remove, this skill refers to what some call mindfulness. Here, I refer to the ability to focus on and feel an experience as it occurs, as opposed to dimming the here and now in favor of a memory of the past, or leaning on our dopaminergic nature to escape the stress of the present in favor of imagining some more pleasant possibility in the future.

This is easier in some situations than in others. When you're having good sex, you're pretty much focused on the here and now without having to try. Other experiences require us to make a significant effort. Even the most committed parent will find it difficult to keep paying attention in the second hour of your kid's junior high choir concert. But by making the effort to experience with our senses what's happening around us, we condition ourselves to become more naturally observant, which better equips us to make decisions about how we wish to react in the moment and live overall. It also gives us the best thing of all: a stronger ability to savor the ineffable qualities that define the experience of being aware and alive.

The last two skills are primarily centered on dopamine-dependent planning.

Skill Five is Value Identification. When we are direct with ourselves about what our values are, we are more likely to align our day-to-day living with our self-conception. That will bring us greater satisfaction with life and lead to more consistent and satisfying feelings and outcomes. Think of value identification as our aspirational side: we may not always live according to our best intentions, but we more or less know what we're aiming for. This is the most dopamine dependent skill of the six: it requires us to manipulate abstractions, imagine various situations in the future, and calculate potential outcomes based on often elaborate models of what might be.

Skill Six is Committed Action. If value identification is a way to point yourself in the direction you wish to go, committed action is what you do to get there. The time for thinking and planning is over. Now it's time for doing things—carrying out your intentions based on the dopamine-driven planning and calculation you worked on before. It is only in this step that you transform those imagined plans into reality. It is also at this step of ACT that highly dopaminergic people may be tempted to rework their plans instead of carrying them out; therefore it's a good time to remember that progress doesn't occur if you let the perfect be the enemy of the good. It's better to take action and adjust your course along the way than to second-guess yourself based on new ideas and information. Better to modify your plan as necessary while you're doing things than to put off getting started.

The goals of ACT and these six skills are to (1) free us from purely reactive behavior, (2) equip us with greater capacity to accommodate barriers that may not be surmountable, and (3) set us on a path of action to reach our goals. In this way we live more closely, and more intentionally, in harmony with the intentions we have for our lives.

A BIGGER IDEA BEHIND IT ALL

There's one more therapeutic philosophy that's worth considering here: *logotherapy*, from the Greek *logos*, referring to "meaning." Logotherapy was conceived in the 1940s by Austrian psychiatrist Viktor Frankl, who asserted that our primary motivation is the need for meaning. Frankl's ideas constitute a therapeutic method and framework by themselves, but they are more often used as an additional layer to enrich other methods. Frankl's ideas can especially enhance ACT, which directs us toward changing the things we can change, most of which are within ourselves, and accepting what we cannot.

Sigmund Freud said we are motivated by the will to seek pleasure. His contemporary, Alfred Adler, who is less well-known but who has had more influence than Freud on modern therapies such as CBT, said we are motived by the will to secure power—the need to overcome feelings of weakness against forces in the world and in ourselves. As these ideas were becoming intellectual pillars of the twentieth century, Frankl proposed something else. On his release from a Holocaust death camp, he reflected on his medical and psychological training in light of his experience there, and on the outlook he and others employed to survive. This led him to an insight on human nature that many consider as vital as the central ideas of Freud or Adler toward a complete portrait of human motivation. Frankl said we are primarily motivated not by pleasure or power but by an innate will to attach meaning to our lives. The addition of Frankl's ideas can be highly effective in informing treatment.

For those independent enough to recognize the value of Frankl during an intellectual era in which Freud held spectacular sway, logotherapy offered a profound advance in therapeutic approach. In particular, he connected an analytical view of the mind with rational valuation of the human tendency toward spirituality and religion.

Frankl widened the path for helping people feel better. If you consider what you know about dopamine, you'll understand why. Freud and Adler believed that we are motivated to pursue things that can never be completely possessed: the stimulation of pleasure and the security of control. We achieve them by attempting to conquer elements of the world—a dopaminergic pursuit that inevitably returns us to a state of discontent. Frankl believed we are motivated by the will to meaning, which lies not simply at the end of dopaminergic pursuit but all along the way.

That's a huge idea with deep potential: *we are motivated by the will to meaning, which lies not at the end of dopaminergic pursuit but all along the way.*

By recognizing that this is how the mind works, we might holistically accommodate and even roll back much of the distress that comes with being alive, including the dissatisfaction that follows the dopaminergic urge. Frankl recognized that what you and I call the "dopamine chase" is never going to provide satisfaction. He quoted Nietzsche's *Twilight of the Idols*: "he who has a why to live for can bear almost any how."

The search for meaning is not a way to distract us from gazing into the void. It provides a way to acknowledge the void and navigate our lives anyway. With that in mind, the therapist's job becomes helping the patient work through the overriding question, *What is the meaning to your life?*

Frankl stressed that the therapist has no right to dictate meaning—it's impossible anyway, he said, since meaning is individualistic, subjective, and re-evaluated over time. A painter, he said, shows us their view of the world—and a therapist shouldn't be like a painter. A therapist's task is more like that of an eye doctor, who wants us to see the world as it is. Logotherapy shows us a way to find achievement in the experience of suffering, and suffering is what brings many to therapy in the first place. Logotherapy helps us find a reason to take constructive action even though life is finite. And it uses our sense of guilt to help us recognize when we're acting against our values; this to encourage us toward a more meaningful journey through life. Taken together, we maintain hope and continue to seek meaning in the face of what Frankl called "tragic optimism," creating personal enrichment against the unavoidable whetstones of pain, death, and guilt.[56]

We tell stories to ourselves and each other, narratives that help us navigate our feelings and problems. A narrative informed by logotherapy can encourage us through a lifetime of challenges, showing us how to find strength through recognizing our own limitations. It can lead us to thrive and discover joy, even in the worst of circumstances.

ANOTHER KIND OF THERAPY . . . BUT IT'S NOT FOR EVERYBODY

I teach a graduate course in public speaking. I give lectures and lead discussions, but the most valuable part is the assignments. Week after week, each student selects a topic, prepares their materials in advance, then has to deliver a six-minute talk in front of a full classroom. You never know when it'll be your turn. When I draw your name, you get one minute to gather your thoughts and set up your slides, then you have to start talking.

It's an intimidating experience. Every semester a few students drop the course on the first night. But for those who stick around, the payoff comes fast. The day after their first presentation, I often receive satisfied messages from my students. A recent text: "I have a terrible fear of speaking in front of a room. A little of that went away after class." Confronting a stressful situation can be the first step to being able to deal with that stress—and also toward gaining control over unwanted behavior.

We can quickly improve our behavior when we face our fears head on, especially when the thing we fear does not pose a physical threat. This is called exposure therapy. One version of exposure therapy involves a gradual approach. If you're afraid of driving over bridges, you could start by crossing a very short bridge, then move to longer and longer ones. The other version is much faster, known as "flooding." In this case, you begin with the most intense version of your fear and experience it completely for an extended time. You could do that by visiting, for instance, Louisiana's Lake Pontchartrain Causeway, which at twenty-four miles is the longest bridge in the United States. If you can cross that one, you can cross them all. In both gradual exposure and flooding, the technique is to experience the fear to gain a reflexive understanding that the danger is imagined, not real, and that the bad feeling is only a feeling, not the reflection of a real threat. You do not attempt to fight it and you do not flee. If a therapist or support person is in attendance, they only observe.

Flooding is stressful and not for everyone, but the payoff is big: it tends to work fast. A few people are freed from their fear after a single session, though booster sessions are typically required or at least helpful, as subsequent efforts reinforce your new capacity. The flooding experience can ultimately reset the body's fight-or-flight system. Once that occurs, a situation that you previously experienced as a fusion of feeling followed by thoughtless reaction is now three experiences: the emotion of exposure, the processing by the observing self, and the calculated reaction. You gain the revelatory understanding that despite your fear, nothing terrible has befallen you. (Remember, we're talking about fears of things that are not dangerous.) You get the further benefit of having this experience while inside the belly of the beast—or rather what you thought was a beast. This profound realization, occurring in a moment of high emotion, can quickly condition your mind to simply remove panic or obsession from the menu of options when you're confronted with the frightening thing in the future.[57] We are thus able, sometimes very quickly, to get rid of the feeling by robbing it of its motivational effect.

A variation of exposure therapy, Cue Exposure Therapy, is often used for addictive behaviors and certain compulsive behaviors, and some people may find it useful for other issues similarly related to dopamine. The idea is to break the link between a troubling cue and an unwanted response. Over time this approach can weaken or outright eliminate the association.

Consider a simple case. Let's say you catch a glimpse of the power light on your Xbox, and that triggers a strong urge to play a video game. You know, however, that your time would be better spent on a project for work that's due in a few hours. Allow yourself to feel the attraction of the activity. Let it truly sweep you up. At this point, you have three things to do.

1. Commit to riding out the feeling, also known as "surfing." This means acknowledging that the feeling will pass, and accepting the unpleasantness of the here-and-now experience by reminding yourself that it is only a feeling and there is no threat.
2. You can more ably do this by carrying out the second task: using coping strategies such as relaxation practices, finding distractions, and asking yourself direct questions about the value of unhelpful or discouraging thoughts.
3. With this much achieved, act on the third task, engaging with the observing self. Exercise mindfulness. Consider what you want in the moment compared to your larger goals for your behavior. You'll think about not just reason and fact but also your aspirations, self-image, and values. In this way you will gain control over a stressful moment by pausing to seriously consider what you want for yourself and how to get it.

In the Xbox example, you'd begin by accepting the unpleasant feeling: *I want to play but I'm going to try to skip it. There's a conflict between what I want now and what I want in the long run. This feels lousy but it's only a feeling. The worst that can happen is I'll feel bummed. There's no danger here.*

Then, use coping strategies to make that frustration easier to accommodate. For instance, try "box breathing": close your eyes, breathe in through your nose to the count of four, hold your breath to the count of four, then breathe out through your mouth to four. Pause for a few seconds, then do this five more times. You might also try progressive muscle relaxation: sit comfortably, then tense and relax your muscles, one area at a time, starting with your toes and moving all the way up to your face. At the end, compare how much less tense your body feels to how it felt when you began. Also note that you likely feel less stressed about the situation.

Now that you've begun to defuse your feelings from your reaction, move to the third task: consider why you're choosing not to play just now. Make a mental list of the benefits of not playing, and what you might achieve instead with the time. Visualize yourself later in the day. You chose not to play. Would you be proud? Happy? Frustrated? Pleased with yourself?

After the feeling from the trigger has passed, you can reflect on the entire experience to write down the things you did that helped you avoid engaging in the unwanted response. You are also entitled to give yourself some encouragement; your effort helped to diminish the connection between the trigger and the unwanted behavior, which will help you avoid that behavior more easily in the future.

In a 2021 study conducted in Australia, researchers worked with a group of individuals who self-identified as having a significant problem with online gambling.[58] To help them resist the urge to use gambling apps on their phones, clinicians gave them tools to prepare before they were in a situation where the urge would strike. Researchers worked with participants individually to devise for each a detailed list of experiences—cues—that made them want to place a bet. In assembling the lists, participants spent time thinking about what did and didn't tempt them, ranking the order of these cues, and considering the experience not just of betting but also of anticipating the act of betting and of waiting on the outcome. Researchers then exposed the participants to the cues and increased their exposure over time. The participants were asked to fully experience the feeling of the unwanted attraction to gamble, and to let themselves experience it until they became habituated to it. Over repeated efforts, all six participants in this small but rigorous study showed improvement in their resistance to engaging in online gambling.

This study, and similar examples in and out of laboratory settings, suggests that we can better push back against unwanted behavior by exposing ourselves to it after preparing well for the experience. That effort begins with identifying the cues that bring on the unwanted urge, then equipping ourselves with strategies to process and reject it.

One especially promising outcome was that it doesn't always take profound sacrifice for this approach to succeed. While all participants in this study wanted to beat their gambling problem, many felt it was unrealistic to stop carrying around a phone—and these participants were more successful in beating the gambling problem than the others. Given that we can't simply eliminate most difficult dopaminergic experiences from life, it's encouraging to know that we may be more successful by learning to cope—wearing down the habit—instead of cutting ourselves off from what troubles us.

CHOOSING AMONG THERAPIES: THE STANDARD IS WHAT HELPS

I was trained as a scientist. Later I turned to writing and the teaching of writing. The combination enabled me to quantify certain types of professional writing so anyone can do it. But my scientific training gave me another conviction: I need an open mind because there are often other approaches that can work just as well as mine. That's true for everything.

If you're interested in therapy, you may find that some approach beyond the ones in this chapter is more appealing or more successful for you. That's good news! What you're looking for is relief and improvement. Finding anything that helps you (that doesn't also hurt you in some other way) is an excellent outcome. A method is validated by its success.

No psychotherapies are direct descendants of completely objective science. Sigmund Freud created psychoanalysis out of his personal experiences and ideas, not from reams of data. Alfred Adler created his therapeutic approach the same way. Freud tied behavior to sexuality and violent instincts outside conscious thought because he thought it made sense to do so. Adler stressed the need to overcome a sense of inferiority to find power as individuals. Neither of these grand theoretical edifices came from anything more than observation and analysis. They're powerful tools, to be sure, but hardly the objective last word.

The frameworks proposed by Freud and Adler are not literal

explanations of how the brain or even the mind functions. They do, however, yield insight when considered as a metaphor, and scientific theory is often most useful as predictive metaphor. In such cases, we are not overly concerned with the method of action. For instance, we can assume the sun revolves around the earth or the earth revolves around the sun. Only one is true, but both provide an accurate forecast of what we will see at sunrise tomorrow. The shocking, seemingly antiscientific conclusion is that if something forecasts the appropriate outcome in a given circumstance, that's good enough.* It may incorrectly identify the method of action, or fail to identify any method at all, but when our goal is forecast, repair, or improvement, a system that gives good results is, for many situations, the only thing we need.

ACT is a strong tool for addressing dopamine-driven challenges, but what's most important is whether it works for you. Consider an extreme example. I have a couple friends who are adherents of alternative approaches to psychology, things that in my understanding have at best a limited foundation in quantitative science. Whenever they bring up something about it, they precede it with, "You probably think this is nonsense, but . . ." I tell them that just because I don't believe in how they say it works doesn't mean it doesn't work. In fact, they've told me many undeniable success stories. I'll never grant that what they're into comports with human biology as we have long understood it. It doesn't. But there are lots of things in recognized medicine that shouldn't work yet correlate with positive outcomes. Most we cannot explain, but over time we have come to understand a few. Consider the placebo effect, in which significant numbers of patients feel better despite being given an inert compound as if it were an effective drug. Recent studies show that a placebo administered within the familiar "ritual of treatment" is especially effective in pain management, stress-related insomnia, and

---

* A friend of mine notes, "Science is always wrong." What he means is that the explanations we have today will be refined tomorrow and that, over time, nearly all of our ideas about how the world works will be discarded. This is why scientific consensus is at once valuable and deadly: it can lead us to useful outcomes but when we treat it like unquestionable writ, we reject science itself, which is no more than another word for rational inquiry.

nausea—conditions that are typically influenced by endorphin, dopamine, and other neurotransmitters.[59] If you're in pain, unable to sleep, or nauseated, your need for relief is all that matters. If what you're trying doesn't work, try something else.

When you're dealing with dopamine-related challenges, or any mental or emotional problem, the best therapy isn't the one that looks good on paper, it's the one that works for you. As you look for the right approach, start with what is most likely to succeed, given what has worked for other people in your situation. When it comes to dopamine-related problems, you will probably find the fastest relief with ACT. If that doesn't work as well as you'd like, consider CBT. If you're still having a problem, consider another practitioner within those disciplines, or look into other approaches.

DISPOSITION VERSUS DISCIPLINE

One last thing about therapy: be honest with yourself about why you're seeking it.

Some problems are diseases or conditions—pathologies—that respond best to formal treatment. Other problems, while unpleasant, are common and occur in passing: modest marital stress, the day-to-day challenge of parenting, frustration with work. To make therapy the default option for nearly every plebeian difficulty is to reject the valuable work of self-discipline toward personal growth. This in turn denies us the opportunity to build the resilience we need to navigate the normal challenges of life.

Seek help when you need it, but think about your own abilities, too, and the resources of friends, family, faith, and experience. Many of us are anxious to pathologize every difficulty, to give it a name that leads directly to some well-known treatment. This makes sense for physical things like heart disease or diabetes, where you either have the condition or you do not. For situations involving difficult behavior and feelings, however, the "diagnosis" can be subjective. Figuring it out is made even more difficult by the fact that claiming certain conditions can gain us sympathy, status, or attention. So ask yourself, are you really dealing with illness that requires intervention, or are you attaching a fashionable name to the problem in order to avoid the hard work of fixing something that is in fact *in your power to change?* In other words, are you correctly diagnosing illness or are you finding a way to avoid what we like least to do: develop greater self-awareness and discipline? *I can't help it, don't you know—I was born this way* is an easier answer than *I need to say no more often.*

Part III

TAMING BIG ISSUES, ONE AT A TIME

To Begin

HOW TO BECOME A LION TAMER

In part II, we investigated approaches to raising and lowering overall dopamine level. In part III, we'll move beyond this general "more or less" problem to take on specific dopamine-related feelings and behaviors plus ways to precisely address them.

Dopamine presents issues (and opportunities) in myriad ways, so we'll need every trick we can find. Imagine you're a lion tamer. A lion is not very good at following directions, which means you need to create incentives that encourage or even "trick" him into doing what you want. You might try to influence the lion's behavior without requiring a change in the lion's nature, since you can't much affect that. So you distract him, direct his energy elsewhere, or calm him. That way you're more likely to get the outcome you want. It's the same with taming dopamine. Most of what we do is to channel its power in more productive or at least nondamaging directions, or, when possible, reduce or elevate its activity in particular pathways in the brain.

Gasoline can power your car but it can also burn down your house, so we have to use it with caution. The same is true for dopamine. It truly is the engine of progress and inspiration but it attracts us to those things in the same way it draws us to risk, dissatisfaction, and danger. Success in minimizing or in some cases eliminating those things is achievable for anyone willing to stick with the effort, but discipline is not the main thing one needs. With respect to the challenges of modern life—dating, sex, social media, obsessive shopping, the lure of gaming—and the inscrutable aspects of joys from creativity to romance, taming the power of dopamine is never a matter of simply gutting it out. Success requires fresh practices, sustained commitment, new ways of thinking, and at least one very old idea, as we'll see in the Finale. Thus we may overcome the downsides of otherwise marvelous modernity, these updated versions of problems as old as mankind. The result can be a happier, more satisfying life. With enough insight and effort, we might thrive beyond our biggest dreams.

Dopamine-driven problems often have overlapping solutions.

For example, challenges like excessive gaming, online shopping, or overuse of social media can be addressed with similar strategies such as tools to limit access or block content and planning alternative activities. This overlap is good news: mastering a few key techniques can help you address multiple problems. Ultimately, understanding their similarities will enable you to more easily recognize and overcome common challenges associated with dopamine.

Still, specific behaviors benefit from a tailored approach so for each I've differentiated the approach you should take based on the particulars of the problem. Many of these helpful approaches have multiple applications, which means you will need to master fewer approaches rather than more.

Chapter 6
WRESTLING WITH ROMANCE

A PHYSICAL THING

For most of history we have consigned the examination of love to the broad strokes of the philosophers, the dreamers, and the participants. Perhaps that's just as well. Comedian Martin Mull said, "Writing about music is like dancing about architecture." The same is surely true of attempting to put love in quantitative terms. There is little concrete scientific study of romance unless you count scribbling down people's feelings and tallying up broken hearts. Neither curiosity about love nor the damage love can leave in its wake—nor the pleasure we take in its sundry delights—yield much insight into its method of action. I quote Paul Rudd's lovelorn character from the movie, *The 40-Year-Old Virgin*: "Love is a mysterious fig."

Only a few things are certain and chief among them is this: we seek romantic love because we are compelled to pursue reproduction. This primary motivation is often buried under a mountain of unrelated feelings, but it is still the main idea. And while love is not completely bound by biology,* it is a terrible mistake to believe that we can overcome entirely whatever nature's part may be. Our experiences result in a feedback loop aimed at modifying our behavior so that we may more successfully pursue romantic love, which arises from "an ancient cocktail of neuropeptides and neurotransmitters."[60] With its role as the insatiable motivator toward the novel and new, dopamine is at the top of the list.

Biology establishes the starting point for behavior because it is the nature of the body that determines what feelings are possible in the first place. For instance, we can't switch off the imperative to survive, though in extreme circumstances we can muscle up the ability to override it for an instant, such as when a parent sacrifices their life for their child's. Biology establishes the range of appetites, appreciations, likes, and dislikes we may know. Experience refines the list but even that is bound up in the physical: we like things that taste, smell, feel, sound, or look "good," depending on the definition of "good" provided to us through our senses, which are biology-bound probes of the physical world. How do our preferences coordinate with biological imperative? Which has the upper hand, why, and when?

Romantic love is a part of the human condition and has been for a long time. For some time, scholars believed that romantic love was a fairly new development, born in the Middle Ages, but a 1992 study still cited today ruled out what anthropologists Ted Fische and William Jankowiak called "an affectionless past."[61] Researchers found that romance is part of being human regardless of where you live, though the experience varies from culture to culture. For instance, people from Asian countries tend to describe love in more negative terms than Westerners, perhaps because of differing priorities placed on social and community obligation. There is also the consideration of the relationship between love and marriage. In countries where arranged marriages are the rule, love is expected to follow marriage (if it occurs at all), not precede it.[62]

---

* Though perhaps it is. "Eventually, as brain imaging is refined . . . it may become obvious to everyone that all we are looking at is a piece of machinery, an analog chemical computer, that processes information from the environment." From "Sorry, But Your Soul Just Died" by Tom Wolfe, *Forbes*, 1996.

Times change, but the nature of a living creature tends not to. That's another reason to respect the role of biology. Understanding love in this context, and understanding the role of dopamine in love, will give us ways to keep relationships vital, better navigate heartbreak, and tame romance.

THE THREE COMPONENTS OF LOVE

Anthropologist Helen Fisher describes three categories that are, in various combinations and sometimes alone, how we experience love of all types.

Lust, or strong sexual desire, is typically a major component of romantic love (as opposed to familial or collegial love), especially early in a relationship. The most important neurotransmitters for lust are testosterone and the estrogens, which regulate sexual function and affect desire for men and for women, respectively.

Attraction, or the desire for connection or further knowledge, is a component of all kinds of love and is at the "heart" of romance. The most important neurotransmitters for this category are norepinephrine, which helps mediate alertness and arousal; and serotonin, which affects a wide range of core functions such as memory, fear, and even body temperature. The other important neurotransmitter here is dopamine, since attraction is an expression of both desire dopamine—urge—and control dopamine, which gives us the capacity to use reason to decide if we should pursue it.

Attachment is the sense of having a connection to another person. Attachment is mediated in part by oxytocin, which is key in pregnancy, childbirth, and nursing, and affects trust and relaxation.

Consider different kinds of love as various combinations of these experiences. I feel attraction and attachment to friends, but I probably do not feel lust. After a few drinks on a lonely night, I may feel lust and attraction toward a stranger in a bar, but probably not attachment. I may feel lust and attachment to a partner in a moment of insecurity, but attraction may have nothing to do with it.

Each aspect can exist alone, too: lust in the form of seeking what you might call not Miss or Mister Right but Miss or Mister Right Now; attachment in a long-term relationship that has become a matter of security and transaction; or attraction in that lifelong ache for the one that got away. When it comes to romantic love, the healthiest combination is all three. Lust and attraction arrive at about the same time, attraction grows the relationship, and attachment keeps you together, with occasional recurring flashes of lust and attraction that may diminish over time.

Why would romantic love follow this form? In fact, why would romantic love appear at all? As with most human behaviors, a consideration of human imperative—in this case, the dopaminergic urge to survive and propagate the species—leads us to a pretty good explanation. Lust can come and go quickly. It isn't picky. Therefore lust can lead us to a quick succession of sexual partners. If the goal is to create more humans, the most efficient way to do that would be to engage sexually with as many partners as possible. Lust achieves that in part via dopamine, which pushes us to the next person by reducing our interest in the one we just encountered: the mystery fades and with it, the person's appeal. But then comes a balancing force toward propagation: attraction. It makes us want to hang around, which is important for childrearing.

Human behavior derives from slow and steady change over millions of years. Modern culture may make it possible for a single parent to raise a child, but the untamed environment that preceded this one was not so welcoming. This is how attraction came to be evolutionarily hardwired, which is why it does not simply vanish under cultural imperative to embrace each new form of purported sexual enlightenment.

Finally, attachment can give us reliability and security in a relationship, and this most important sustaining factor can overcome lust and attraction, even over the long term. Put these three qualities together and the species propagates vigorously.

This suggests a step-by-step way to tame dopamine in situations where romance or its possibilities leave us overwhelmed. Dopamine tells us that this new and unfamiliar partner, or this feeling that we're chasing, is exactly what we want and need, but it has no basis for the claim beyond hope generated by unfamiliarity. We need a way to separate what we want from what we need, and what we hope to be true from what is.

ADJUSTING THE INGREDIENTS

To take some control over the situation in terms of the effects of dopamine in lust, attraction, and attachment, let's investigate what we're up against and, from the results, make an informed choice at each point in the process. First, we'll examine and write down the particulars of the situation. This will reveal the influence of the desire dopamine system versus the control dopamine system, the better to recognize when we are overvaluing feeling and not doing enough thinking. Then, we'll give our strategy a better chance to succeed by letting time pass. The hours we spend analyzing the situation will allow desire dopamine to diminish, which can decrease the distorting allure we may be feeling while allowing control dopamine to produce a greater number of reasoned analyses.

Taming Lust. Let's consider lust in the form it most frequently bothers people: a desire whose intensity makes you feel less in control of yourself. This most likely presents as an urge for sex beyond a committed relationship, an unwanted desire for sex without attraction or attachment, or a sex drive significantly greater than your partner's or what either of you might consider normal.

- *List the reasons you don't want to give in.* Do you aspire to limit sex to marriage or a committed relationship? Would this act violate your religious beliefs or ethics? Are you concerned about disease? Do you dislike the way sex in this context makes you feel about yourself? Reasoning your way through your feelings can reinforce your commitment to the decision you want to make.
- *Distract yourself.* Take your mind off the feeling with something you greatly enjoy. Walk your pet, get some exercise, call a

friend, or do something creative such as playing a musical instrument or pursuing some artistic skill.
- *Reach out to a more appropriate object of your affection.* If the feeling involves someone outside your relationship, reach out to the person you're committed to. Call them or go see them in person. Their presence will remind you why you want to limit your intimacy to them alone.
- *Practice mindfulness or meditation.* This can be as simple as breathing exercises or as involved as a focused practice. Think of it as slowing down what may feel like a race to get what you want but in truth do not need.
- *Try elements of ACT.* Ignoring unwanted thoughts doesn't make them go away, so try experiencing the feeling wholly. As the desire washes over you, observe that you are not acting on it. Tell yourself what's true: life includes unwanted and even intrusive thoughts, but exposure to them over time can make them less powerful, and can build your ability to feel them without acting on them.

Taming Attraction. It's difficult to distinguish attraction from a simple reaction to some other feeling such as loneliness. A good metaphor is the desire we sometimes have for food. We eat for nourishment and stop when we're full. That's a valid reason to eat. Other times we eat not because we need food but as a distraction from boredom, and over time that kind of eating can cause problems.

To navigate the problem of attraction, start by thinking of it as an active desire, a positive thing that creates additional or enhanced good feelings as opposed to a feeling that pushes back against discomfort. Then follow these points:

- *Ask yourself what exactly is attractive about the person.* Write down what you like about him or her. Be specific about what initially captured your attention and what has held it. Was it their looks? Their sense of humor? Their fashion sense? The things they have in common with you? The way they make you see

things in new ways? Don't settle for vagaries. Is the attraction mostly physical, sexual, emotional, or intellectual? Really think about those distinctions. The more reasons you can specify across multiple areas, the more likely your attraction is real.
- *Compare the qualities you've written down to those of people you've been attracted to before.* This can help you decide if you're simply pursuing a "type" or if this is someone you're really connecting with.
- *Make note of how you feel when they're not around.* Do you still think of them when you're apart, or is the attraction primarily or even only a factor when you're around each other?

Taming Attachment. We can be attached to someone outside of love. Witness a marriage maintained "for the children" or one based on shared financial needs. A healthy attachment in the context of romantic love is having or cultivating a vigorous connection—emotional and intellectual—with someone else and, typically, sensing that connection and effort in them as well. It includes a sense of obligation and commitment to the other person. It also features a sense of confidence in their obligation and commitment to you.

Attachment is healthy and enjoyable. If the feelings associated with attachment aren't being returned, however, your attachment is likely more from need or insecurity than from love. Review your attraction with these exercises:

- *Consider less how you feel about the other person and more about how that person makes you feel about yourself.* Do you feel secure and connected? Does this feeling persist in a way that is only minimally dependent on circumstances? Attachment in the context of a loving relationship gives one a sense of completeness about purpose and even personality.
- *Consider the consistency of the attachment you feel.* Healthy attachment is not reliant on impulse or passing emotions. Attachment that's worth your commitment reflects commitment itself. It is as much a decision as it is a feeling.

The next time you're angry or offended in your relationship, ask yourself if you still feel or want to feel attached. Assuming you do, ask if it's because you are confident that particular person will make you happier in the long run, or if the attachment is more about the fear of not having attachment in your life at all.

- *Analyze your sense of attachment when you are apart.* It's natural to feel temptations from outside a relationship. Attachment helps us maintain the commitment we've made despite them. If your sense of attachment tends to fade when you're apart, think about why that might be. What feelings in particular are fading? Does your sense of personal completeness fade too? What about your desire for or commitment to shared plans and obligations? If the benefits of this particular attachment lose some of their appeal when you're apart, it's possible that this relationship is more about managing goals and discomfort than maintaining constructive attachment to the person.

As disorienting as romance can feel, we can regain some control by thinking of the experience less as a single feeling and more as a competition among feelings of lust, attraction, and attachment. We can reason our way through each of these, but to succeed we'll need to do more. We'll want to calm ourselves amid the intensity with distraction techniques and mindfulness. We'll also want to pull back on using the observing self to defuse response from feeling. We'll want to analyze what our choices really are. And we will need to imagine their various outcomes using the dopamine-powered capacity of mental time travel.

Chapter 7

FINDING THE RIGHT PERSON

DOPAMINE AND "THE NIGHT WE MET"

Dopamine and the search for "the right one" are knotted up like last year's Christmas lights. There is no more intimately human experience more dependent on both our desire and control circuits than the purposeful search for (and occasionally unwanted temptation of) a potential romantic connection.

This is an occasion when dopamine exerts several of its most impressive powers. The first moments when we meet someone are an opportunity for reward prediction error—we hope for a romantic partner more often than we feel the spark in some first meeting, so an enticing connection exceeds our expectations and fires off dopamine. The desire circuit alerts us to promising possibilities and pulls our attention along. The control circuit explores the endlessly delightful possibilities of being with this person—who so far is about 1% reality and 99% glamorous, dopaminergic imagining of what could be. Consider how seeing a plane in the sky suggests a glorious vacation, but being seated on the plane gives us nothing but a desire to be back on the ground; this resembles our feelings around new potential partners. The most promising part, the dopamine hit of the experience, comes not when the relationship becomes reality, but before. Hence the dopamine-driven challenges around finding the *right* person. We have to remember that what we imagine before us is likely not what awaits.

IS THERE REALLY LOVE AT FIRST SIGHT?

If you're fortunate enough to have the experience, it's the height of joy to feel "love at first sight" as it happens—or seems to. Months or years later, it's almost as delightful to look back and remember—or simply believe—that love really was there from the start. But does love at first sight actually happen? The feeling is real, but it may be no more than that—a feeling. Consider what love really means to you, then ask if love at first sight is possible. That's quite a lot of commitment to hang on a glance and a smile.

Are we just romanticizing the here-and-now moment we crave—the beginning we'd prefer?

Love at first sight is related to the actions of several neurotransmitters. Oxytocin, the famous "cuddle" hormone, would seem to be the place to start but it's not. Oxytocin is less important in the feeling because it's more related to bonding and trust over time than to an initial spark of attraction. Serotonin matters not because it increases but because it decreases, perhaps as a response to the stress of a new encounter, and reduced serotonin can contribute to having obsessive thoughts about a person. Norepinephrine gives us the alertness and focus we feel in that moment, and endorphin gives us pleasure, even euphoria. But it is the surge of dopamine that is most important, coming at the start in an episode of reward prediction error. The mismatch between our expectations and our excitement about whatever it is we find in this new person—that's what very well may kickstart the feeling.

FINDING THE RIGHT PERSON

A European study of men and women in their mid-twenties collected information on LAFS* and the experience of meeting potential romantic partners. In addition to answering questions, participants met up in "dating events," where researchers collected "hot takes"—responses about how participants felt in the moment—instead of giving them time to reflect and potentially swap what they really felt when meeting others with what they decided about them later.

Researchers asked participants if they felt LAFS during an initial meeting. Then they asked other questions, one of which turned out to be especially significant: how physically attractive the participant found the other person to be. When people said they felt LAFS, it was usually with someone they considered especially good looking. There was no similar correlation between LAFS and high initial ratings for "intimacy" and "commitment."

Driving home the stereotype that men fall in love with their eyes, men were more likely than women to report literal love at first sight, meaning as soon as the man saw the other person. For what it's worth, in none of these cases did the object of affection also report the LAFS experience. In Gilbert and Sullivan's *Iolanthe*, Lord Chancellor sings, "Love, unrequited, robs me of my rest." The chancellor seems to have plenty of company.

The study authors concluded that at its root LAFS is a romanticized characterization of "strong initial attraction." It appears to be more about liking what you see rather than the product of some extrasensory tell or unconscious processing. Which is not to say it doesn't have its utility. Though the researchers characterize it as what they are too kind to call self-delusion—their Willy Wonka–style term is a "memory confabulation"—they do grant that LAFS can have a purpose. Perhaps, they say, we talk ourselves into believing in it as a way of enhancing our relationship. Savor the irony: We claim LAFS when we kick off a relationship as often as we embrace it to stay together.

---

* Researchers abbreviate "love at first sight" as LAFS. (Say it out loud.) Maybe they're being ironic.

Real or imagined, love at first sight is a delightful feeling. If you simply want to experience love at first sight and you don't care if it's mutual, it's pretty easy to do: since physical appearance is the primary trigger, find somebody you think is hot, then conjure up LAFS on your own.[63] Who knows? The other person may even reciprocate. Enthusiasm goes a long way.

CAN YOU MAKE SOMEONE FALL IN LOVE WITH YOU?

The surprising answer: You have more power than you think. An even bigger surprise: Using a simple approach that puts honesty at the forefront, you can pick just about anybody and have a pretty good chance to connect—with one caveat, and it's a big one: they have to be as willing and honest as you are in sharing themselves.

In 1997, five US scientists presented a "practical methodology . . . for creating closeness in an experimental context."[64] They started with the idea that "closeness" is an intermediate step toward love, meaning two people coming to know things not typically associated with romance: facts about each other's lives, personal vulnerabilities, and individual perspectives on the world. Here's the important part: The two people have to use this information not to form a judgment about the other person, but simply to come to know them. With that stipulation, the goal of the experiment was simple. The researchers suspected that if two people got to know each other by incrementally sharing increasingly personal things, the exchange might cultivate not just the feeling of closeness but actual closeness—and, perhaps, love at first sight. They chose to carry this out in a rather clinical environment, too, the better to rule out the effects of typical new-relationship variables such as booze and loneliness. They wrote out a model for conversation to expedite this kind of exchange and gave it to the couples to follow. (The *New York Times* published the questions in a 2015 article. If you'd like to use them yourself, see the endnote.[65])

In the experiment, each couple spent most of an hour together asking each other questions in fifteen-minute sessions. In the first session, they focused on basics, asking about personal likes and dislikes, life goals, and self-description—icebreakers to establish rapport. The second session got a little more personal, requiring vulnerability and honest self-assessment. Questions touched on significant memories and accomplishments, regrets, core values in friendship, family, and love, and the occasional foray into hypothetical situations—"What would you do if someone asked you to . . ." kinds of questions that reveal personal priorities. In the final session, the dialogue was crafted to build intimacy and create connection. The couples asked each other to share problems, to talk about feelings and personal issues, and to open up a little about their significant fears. On a happier note, they were also asked to identify positive qualities they now saw in each other.

After the conversations, most participants said they really did feel some closeness to their new partner. The study showed that many of the things we assume are critical to feeling close to someone don't matter all that much. You don't need similar attitudes about world events or matters of taste. You can come into a conversation with expectations about a person's likes and dislikes, find out you were wrong, and, at the end, still feel close. You can even state up front that the conversation is calculated to make the two of you closer and this will make it neither more nor less likely that you'll end up closer. It turns out that the key factor to closeness is sustained and shared self-disclosure, honesty, and vulnerability. Common outlook and attitude may matter greatly in later phases, but when it comes to simply "hitting it off," none of that matters. So it may in fact be possible to induce a little love—if not at first sight, then perhaps by the end of forty-five minutes.

While a calculated conversation over coffee isn't as memorable as a thunderclap of instant attraction, it's pretty amazing to find that a form of love at first sight (or at least at first conversation) is not only possible but also can be created—with some effort and a little cooperation. You don't get the fireworks, but you do get a significantly better chance at a connection. (In the study, which was carried out in a college psychology class, one of the couples eventually married!)

The role of dopamine in the process points toward lessons that can help us in any situation where we're trying to develop a romantic bond.

Deep conversation can be powerful. Dopamine is triggered by the unusual, novel, and new. We don't often share our vulnerabilities and deeply personal experiences, let alone with people we're only getting to know, so the experience of doing so—when it's reasonable to do so—can trigger dopamine and its attendant good feelings. That can make a conversation more enjoyable and memorable, and those qualities encourage attachment.

Social reward breeds more social reward. Human beings are inherently social. When a social interaction goes well, we feel good about it. When we secure the social reward of connection we will tend to seek and find further interaction.

Having a goal encourages reaching the goal. The prospect of creating a meaningful connection with another person, especially a stranger, is a well-defined goal, and having such a goal triggers the release of dopamine. That in turn motivates more goal-driven behavior.

When we go looking for a date, we focus on appearance and clever lines. If all you want is a dance, a drink, or a night together, that can work. But if you're looking for a relationship—an honest-to-goodness love connection—one way to kickstart the opportunity is to really get to know someone by being open, honest, vulnerable, and engaged. That doesn't mean rushing in with a bunch of confessions and personal details. It just means taking a measured approach toward revealing yourself and learning about the other person. The only caveat is that they have to be willing to do the same. You don't have to make it weird. Just start the conversation with light things then steer it toward matters that are deeper. Take your time. It works better than you may have imagined.

WHAT REALLY HAPPENS AT CLOSING TIME

In 1976, honky-tonk singer and saloon proprietor Mickey Gilley was one of the biggest names in country music. That January, he had his fifth number-one country single on the *Billboard* charts, "Don't the

FINDING THE RIGHT PERSON

Girls All Get Prettier at Closing Time."* This bit of trivia matters to us because it mattered to Dr. James W. Pennebaker, who in 1979 was a newly minted assistant professor of psychology at the University of Virginia. Intrigued by the song's flippant titular question and common sentiment, Pennebaker and colleagues undertook a legitimate investigation, exploring the question at hand from the perspective of both men and women (though he limited his study to heterosexual attraction), and published their findings in the peer-reviewed journal *Personality and Social Psychology Bulletin*. The research, which gave birth to the term "closing time effect," was serious, though the tone and the language are pretty funny:

> *Despite psychology's attempts at keeping pace with hypotheses generated by song writers, research dealing with perceived physical attraction has fallen far behind. In an attempt to close the gap, a study was conducted which confirmed Gilley's (1975) prediction that "all the girls get prettier at closing time, they all get to look like movie stars . . ." A* reactance *[emphasis added] interpretation based on pre-decisional preferences validated Gilley's observation "ain't it funny, ain't it strange, the way a man's opinions change when he starts to face that lonely night."*[66]

Think of how you feel when your boss tells you to do something. Your first reaction, at least in your head, is often, "I'll do it when I get around to it!" or, "Go ask somebody else!" This is *reactance*, and it can lead us to be the opposite of our best self. The relationship between reactance and dopamine is not fully understood, but we can think of it as the result of a dopamine surge triggered by threat, in this case the threat to independence and autonomy. A second of reflection can restore our judgment, but our go-to is often, "Don't tell me what to do." It can be such a strong impulse that we prioritize pushing back over our own standards and best longer-term interests. Threaten even a sliver of

---

* This observation may be true about boys as well, of course, but this was 1976. Also, putting all that in there would have wrecked songwriter Baker Knight's rhyme, though it probably didn't occur to him to begin with.

our autonomy, and that autonomy becomes the most important thing. For someone looking for an immediate sex partner, closing time means all the choices are about to disappear. That's a threat to autonomy. We may have a reactance reaction that lowers our standards.

Pennebaker also wrote that an alternative explanation comes from dissonance theory. If you're committed to "picking up" someone at a bar, you favor people you consider attractive, but such people may not be around. To resolve the difference between what you want and what's available, you might convince yourself that the people around you are more attractive than you first believed. He cited a "classic finding" from a 1970 study: "As time to make a decision diminishes, the perceived attractiveness of the alternatives converge."[67] The research compared assessments of attractiveness if you have to settle on someone in fifteen minutes versus three minutes. The finding: The more time you have, the more you distinguish between people being more or less attractive, but as the time you have to choose goes away, you tend to raise your earlier assessment of a person's attractiveness—a wordy paraphrase of what Mr. Gilley knocked out in a dozen syllables.

Pennebaker's research method put all this into play. He sent three male–female pairs of survey-takers into three bars at three times: three-and-a-half hours from closing, two hours from closing, and thirty minutes from closing. Men interviewed men and women interviewed women. The question was simple: "On a scale from 1 to 10, where 1 indicates 'not attractive,' 5 indicates 'average,' and 10 indicates 'extremely attractive,' how would you rate the [people of the opposite sex] here tonight?" Then the questioners asked about members of the target's own sex. The result: While same-sex ratings remained the same, *opposite-sex ratings increased as closing time approached.* The work not only validated Gilley's "hypothesis," it also gave this old bit of folk wisdom a grounding in psychology.*

---

* If there is an oversight in Pennebaker's scholarship, it is that the team focused solely on Gilley's assertion while skipping over a similar pioneering claim made five years prior. Singer-songwriter Stephen Stills blazed a trail all the way to number fourteen on the Billboard Hot 100, complete with a splash of literary antimetabole. His antidote to our inability to be with the one we love? Love the one you're with.

How to resist reactance or dissonance? As with so many behavioral issues, the first step is becoming aware of both the problem and its source, and you've already done that by reading this section. When you catch yourself pushing back against a reasonable request, whether from someone else or from your own higher priorities, recognize that it's happening. Instead of clearly assessing the situation and then making a thoughtful decision, you may be pushing back against a threat to your independence, lying to yourself about the value of something, or both. Either way, you're trading the truth to get what you want. You can choose to do that; just be sure you're aware that it's a choice. You may want to choose something else.

Try this: Tell yourself not that you "have to" do this or that but that you "get to." This isn't easy, at least not the first time. Doing it is going to feel forced to you—because it is. But that doesn't invalidate the approach. When you change your language from "must do" to "choose to," you are reclaiming the authority and autonomy you forgot you had and rejecting reactance. You're no longer the one taking orders. You're the one giving them. For instance, *If I don't want to spend the night alone, I have to find someone as desperate as I am* becomes, *I'd like to be with somebody tonight, so now I get to decide who I might ask. Who is worth my attention? Who does or doesn't measure up? Do I want the possibility of sex more than I want to stick to my standards? Or are my standards worth something more? What do I want to do, really?* When you talk to yourself this way, you choose to disengage with reaction in favor of making a reasoned decision. You're back in charge, and you're considering your choices as they are, not as you wish they would be.

THE APPEAL OF DANGEROUS MEN

Is the "appeal of dangerous men" a toxic stereotype or a fact of life? For some it's a touchy question. What may be a trend in a group is not true for every individual. No blanket statement in this realm applies to everyone.

With that caveat, here's what some of the most recent data seems to tell us. A 2023 study from researchers in Australia suggests what you may have suspected: women seeking short-term relationships or brief sexual liaisons tend to be less attracted to "careful men" and more attracted to those who engage in risky behavior—"dangerous men."

"For casual sexual liaisons, women prefer courageous 'cads' with a good genetic constitution—risk-takers seem to fit this bill," said Dr. Cyril Grueter, the study's lead author.[68] If we think of relationships in transactional terms, this is what we would expect to find. The appeal of a short-term connection begins with enticement, the potential for a better-than-expected experience, the feeling caused by triggering the desire dopamine system. What more obvious source of the unexpected can there be than someone who announces his risk-taking with his words and behavior? The transaction also comes with a bonus: the "risky" experience of dating a dangerous man is likely to be pretty safe for the woman, since it's the other person taking demonstrative chances. This gives the control dopamine system a strong reason to endorse the decision.

A follow-on finding in the study makes sense in the same way. Grueter and colleagues found that when women are in relatively good health and have easier access to health care, they are more likely to be attracted to risk-taking men, especially compared to women with less economic security. The reason may be that their financial position and the social safety net make it easier for them to deal with the consequences if their choice goes wrong. Think of it as insurance: even if the risk-taking connection leads to disease, injury, or unwanted pregnancy, they can probably restore their lives to what they were before.

The effect of risk-taking on attraction is sexually dimorphic, meaning that there are differences between male and female response. Males are more prone to risk-taking than females.[69] In general, however much risk a woman perceives in a relationship, a man will perceive less.[70] So, the attraction of risk-taking behaviors is reinforced not only by immediate dopamine phenomena but also by a long-term expression of the dopaminergic urge, the evolutionary imperative to propagate the species. Women have profound power to choose who they will engage with sexually. Risk-taking makes a particular man more attractive than another. It's what the study authors call a "sexual advertisement"[71] of the ability to tame the physical world. To that end, the authors note that other studies[72] indicate that men are more rather than less likely to take physical chances in front of women, "especially in the presence of attractive females."[73]

Why is this risk-taking preference among women focused on short-term and not long-term relationships? Because the purposes of those relationships are different. In the short term, the goal is to secure a partner with genetic superiority; earlier studies indicate that "facial masculinity" has a similar appeal to risk-taking and is perceived as an indicator of, among other things, good health.[74] In the long term, the goal is to secure a partner who exhibits valuable behaviors—one who is likely to stick around, be able to make a living, and figure out problems as they occur. This is where evolution promotes certain qualities and demotes others.

Before modern society, it would have been nearly impossible to raise a vulnerable infant while simultaneously hunting for game to provide sustenance and fending off, say, wolves. The world has changed, yet the fact remains: biologically driven forces inform the choices we make in our relationships, and they cannot be disregarded on the observation that they are culturally out of date. Millions of years of trial and error have brought us to this point in how we propagate the species. Nurture lags nature, and it will take a while before our genes catch up with our most recent aspirational behaviors and purported better angels.

In the end, we're entitled to like who we like, and that includes limiting ourselves to safety-seekers, taking a flyer on a risk-taker, and presenting ourselves to appeal to the sort of person we seek. No one owes anyone an apology or an explanation for their romantic attractions.

For women. A woman who is concerned that she is susceptible to risk-takers can better resist by being aware of the reasons she might feel the attraction, then calling on that thinking when the moment comes. Desire-circuit feeling is answered by control-circuit analysis, and we feel far more control in the latter than the former. A woman who considers attraction to risk-takers a failure of character should also remember that it's not a matter of character in any way. It's an expression of genetics, which in this case imposes a steep discount on an individual's more noble characteristics.

For men. A man who wants to be more attractive for short-term relationships and casual sex might improve his appeal with the obvious move of being more of a risk-taker in front of women. This doesn't have to be feats of physical strength. It can be as simple as putting confidence behind his words and saying what's on his mind. Confidence—that's what's behind the appeal, and it can be built with practice. Imperfect attempts and even outright failures help inure us to embarrassment. That makes anyone more confident the next time, which makes a person less likely to fail and less hesitant to take risks in the future.

ARE THERE CLEAR-CUT WAYS TO RECOGNIZE THE RIGHT ONE?

One of the ways we tame dopamine when we're in love occurs when we decide that we really are in love.

Making a purposeful decision that you are in love shifts your focus away from chasing a dopamine hit of new love and the mysteries of a new partner, and toward growing your attachment to that person. This is not only desirable, it's also a constructive way to conclude a dopamine-driven pursuit. Remember, the evolutionary purpose of dopamine is not to give us pleasure in the pursuit as an end in itself. Sometimes the dopamine "promise" comes true. In that case it leads us to pleasure, satisfaction, and security that we experience in the here and now. At this point we switch from the search for love, which is driven by anticipatory, dopamine-powered feelings, to enjoying the experience of being in love, which is governed by the H&N neurotransmitters.

To make this important switch, we need a way to know that we're in love in the first place. We pour out a lot of hours trying to figure out if we're there. The problem is that "am I or am I not?" is so subjective. It's a matter of how valued we feel, how attracted we are to the other person, how well our plans for the future comport with the other person's, and how much we trust our assessments of their feelings and ours.

LiveScience reporters Robin Nixon Pompa and Patrick Pester surveyed the scientific literature and constructed a dozen simple tests[75] to help someone decide if the relationship they're in really is love. While this will never be an entirely objective question, Pompa and Pester's questions give us a methodical approach that can put more confidence behind optimism.

Consider the list of questions below. If you answer "yes" to more than a few of these, it's reasonable to believe you are in love right now and can turn away from wanting love and toward building it.

- Do you think this person is more special than others you've dated?
- Do you focus on positive things about the person more than the negative?
- Do you experience more frequent than normal swings in energy and mood?
- Do you find thoughts of the other person significantly intruding into your time?
- Do you feel dependency in the form of separation anxiety and fear of rejection?
- Do you spend significant time imagining a future with this person?
- Are you more sympathetic to the person's pain than you are to the pain of others?
- Have you reordered basic priorities and even personal interests to better align with the other person?
- Do you find yourself feeling possessive of the other person?
- Do you crave not just sexual intimacy but also emotional intimacy?
- Do you suspect that your feelings for the other person are at times beyond your control?[76]

Finding the right person is a desire-dopamine-driven exercise that begins with casting about for someone different enough to pique our curiosity but familiar enough to remind us of what we value most. Or, to put romance in clinical terms, we begin by looking to be surprised via reward prediction error. A better-than-expected encounter will push us to learn more about the person, but for a while we luxuriate in knowing only a little. We imagine endless and alluring possibilities with them, based on barely anything, hoping for that human connection by visualizing its ideal. Eventually, imagining has to give way to learning, so control dopamine comes in next to fill in the blanks about this stranger. We usually figure out quickly whether this person is a possibility worth pursuing. When they aren't, we lick our wounds and move along. When they are, it makes that potential partner even more attractive—and we continue down the path. Our initial imaginings are transformed into a first draft of what's possible. It is this tentative, titillating, and mutually soul-baring process that defines the search for "the right one."

Chapter 8
MANAGING SEXUAL FEELINGS

SEX AND THE POWER OF COMPETITION

Among all human urges, sex is the most influential on the self, most prominent in the culture, most short-term demanding, most long-term affecting, most likely to lead to embarrassment and loss of status, most quickly connected to impulsive violence, most often the source of brief, intense pleasure followed by a lifetime of trouble, most often the reason to lie, and the most powerful motivator for taking action that can thrill us or wreck our lives completely. When it comes to influence, sex beats the stuffing out of cool reason. It is said that Helen of Troy was so attractive that her face launched a thousand ships. Good for her; sexual desire is a more direct way to get that done than negotiating contracts with dockyards and sailors. Before the rise of equal rights for women, the differences in sexual attitudes between men and women were a primary source of influence and outright power for women. These differences are also why women can have such a civilizing influence on men.

We can tame, mitigate, amplify, and suppress all our human urges to some extent, but sex is the most indomitable. Dopamine is the molecule of more, but sex is the urge for more—and more, and more—and this is part of its uniqueness. Why don't other natural drives urge us with the same intensity toward such extremes? Why does the sex drive push us so far beyond its purpose of self-propagation?

Because an element of sexual "success" is built on competition.

Most drives push us to seek things we can get on our own, like food, but propagating the species takes two. Think of a sexual "opportunity" as a man auditioning for the prize of a woman's attention. When it comes to sexual appeal, if a man succeeds in one such "competition," he's likely to succeed in others. That means not just this particular man but this kind of man in general—a more sexually attractive man, who is also more sexually oriented—will reproduce more often. He doesn't have to be doing anything on purpose to make this happen. It could just be in his nature. Doesn't matter. In the end, this type of man naturally has a greater presence in the reproductive pool.

The history of humankind is generation after generation of the most sexually active and sexually motivated men and women being naturally elevated. Thus in the modern age, the sex drive has become freakishly huge, especially among men. We are saddled with the legacy of our sex-focused ancestors. That's partly why our culture is saturated with casual sexual references. We are more focused on sex than perhaps we'd like to be: we are the descendants of our most sexually fluent ancestors down through time.

All of which is why the most powerful neurotransmitter, dopamine, and its ability to draw us to romance and guide us through the challenge of relationships, is integral to the most powerful biological imperative: survival via the propagation of the species.

IS THERE A PROBLEM, OR AM I JUST REALLY INTO SEX?

Let's be clear: being deeply interested in sex, even obsessed, isn't the same as being addicted. It's likely not even a problem. A deep and

MANAGING SEXUAL FEELINGS

abiding interest in something becomes a problem when it becomes a compulsion—when you can't stop doing it even when you try over and over again, or when it interferes with your job, your education, your relationships, or your peace of mind.

I like pizza. Once in a while I crave it and I'll ask my wife if she wants me to bring home a pizza for dinner, and if she doesn't I'll swing by 7-Eleven and pick up a slice for myself. Sometimes I eat more pizza than I should. But I have never lacked the ability to say, "No, I'm not going to have pizza today." I've never missed work in favor of sitting in the Domino's parking lot serially ordering extra-large super-supreme pies.*

In fact, we can be "addicted" (in the loosest sense of that term) to things that are good for us such as exercise, meditation, or prayer, and we wouldn't want to change it if we were. One may be "addicted" to exercise to the point where you sacrifice other activities in your day to pursue it, but unless you're neglecting your kids or skipping work, that's a good problem to have.

If you're concerned you might have a problem with sex that could be helped by getting professional support, take a dispassionate look at how sex affects your life and consider these questions. The answers may suggest that you should find support from a credentialed expert, or they may guide you toward things you can change on your own or with the help of a trusted friend.

Am I consistently unable to resist the urge to seek sex with a partner? If you ask your partner to share an intimate experience with you more than they would like to be asked, that may irritate them and frustrate you, but on its own it doesn't indicate a problem beyond insensitivity. It becomes trouble when you're asking so often (whether you get turned down or not) that you're interfering with the other person's life, you're not meeting your other obligations, or you're so distracted by the pursuit of sex that it regularly interferes with your day. Still, that's worth addressing too. Being a jerk isn't a disease and fortunately there's a cure: quit blaming it on your biology and stop being a jerk.

---

* See the "pizza" discussion in chapter 4. There's a difference between indulging a craving and being unable to resist it.

Am I unable to resist the urge for sexual activity by myself? With this question we arrive at the far more compelling concern bubbling under the surface of this chapter: *Am I normal?* or, in this case, *Do I masturbate more or less than everybody else?* We finally have data on the matter. Among men and women aged eighteen to twenty-four, about one in four masturbate several times a month. In the same age bracket, about one in ten women do it more often, several times a week. About four times as many men as women masturbate that frequently. There's not much of a drop-off in masturbating multiple times a week until past age fifty, and that's true for men and women.[77] In broad terms, men masturbate more often than women but not by much.

While we might like to know how "normal" we are compared to other people, what matters more is whether solo sex play interferes with our ability to do other things that are important to us. If frequent masturbation is cutting into the time you'd prefer to spend on things you value more, or if it is becoming a substitute for the work you'd otherwise put into seeking or maintaining a relationship, this may indicate a problem in need of attention.

Is my sexual behavior so consuming that I have to lie about it? A sexual liaison that makes you late for a business meeting might be something to brag about, but if it happens so often that you always need excuses at the ready, it calls into question whether you are in control of your sexual behavior. When the cover-up becomes a regular element of the sexual behavior, there's likely a problem you need to address, and the longer you put it off the more difficult the clean-up will be. This is also directly connected to the issues addressed in the next question.

Am I compromising my moral code? While moral compromise is not necessarily part of addiction, thinking about the morality of what you're doing versus the morality to which you aspire can help you more clearly understand the role of sex in your life. It can also provide motivation for the behavior you prefer. If you're in a relationship, you likely share a basic moral or ethical sensibility with that other person. Is what you do a violation of your moral code? Does this violation have a cost? If it does, are you willing to live with that? Is what you do a violation of your commitment to your partner? If they were aware of it, would this be a problem for them or for you?

Do I face potentially damaging financial, legal, and medical consequences because of my sexual activity? When you're dealing with the consequences of sexual behavior, it's clarifying to spend some time considering it as a physical act apart from feelings. Do your actions consume resources that you share with someone else? Pornography can cost money; random hookups, even more. So does prostitution, which also comes with legal and reputational risks as well as hazards to your health and that of your partner and the sex worker. Set aside the moral questions and consider whether what you're doing could have an impact on your profit and loss in terms of money, safety, health, reputation, and professional standing.

Am I using sex to avoid dealing with something else? Some people use drugs or alcohol to lessen stress or avoid facing a painful problem. Is there an ongoing difficult issue that, when it comes to mind, is usually followed by engaging in some sexual pursuit?

Having concerns about your attitude toward sex is not the same as having a sex-related problem that can profoundly damage your life. Sex addiction is different yet again: it is an urge toward sexual activity so strong that it is beyond your capacity to resist, often the result of a dopamine system warped out of normal order. If the questions here set off some alarms, consider seeking help. Think of this section as a screener to point you to help if you need it. It's also here to give you some comfort that you're "normal."

Sex is a dopamine-driven pursuit that gives way to the influence of other neurotransmitters to ultimately produce a profound emotional experience in the here and now. The ideas here are about dealing with being more interested in sex than you'd like to be, being less able to control your sexual desire than you'd like to be, diminishing that interest successfully, and revving it up when it's fading. Some of our priorities are inevitably wound around cultural expectations, pride, guilt, and preference. The idea here is to unpack some of that so we can have the sex life we want.

FRUSTRATION ATTRACTION

The old saw about the "biggest" sex organ is true: it's the brain. Brains can turn the ordinary into the inviting and the weird into the wonderful. This is especially valuable when the excitement of anticipation begins to run short, a common affliction of long-term relationships.

To help sustain a sexual connection, consider a dopamine-driven approach of self-directed denial. In a *Psychology Today* article,[78] Gregg Levoy observed that a fundamental component of passion is the feeling that comes only from being apart, which feeds the anticipation of being together again. Think of it as a carnal take on Juliet's sweet sorrow: we can create delicious sexual anticipation from waiting—call it "frustration attraction." So deny yourself a bit, because the most direct path to sexual renewal is delaying gratification.

To restore sex's intensity, you can engage a therapist, buy a box of toys, or strain your imagination, but none of that is necessary to restore at least some of it. Citing ideas from psychotherapist Esther Perel and Dr. Gina Ogden, Levoy presented several ways to keep passion alive when dopamine has downregulated what it takes to feel excitement. Here are a few of those ideas reoriented to improve sexual intimacy by stretching out anticipation.

The silent treatment. Set aside some period—hours, even a full day—during which you and your partner will not speak to each other. You can do anything else: write, gesture, touch, draw pictures, point—just no talking. You may want to celebrate at the end with an intimate encounter, also free from speaking.

The cutoff. Take the "silent treatment" a step further. Not only will you forbid speaking, you'll also forbid all communication. During this time, note how frequently your thoughts turn to the other person, and why. Do you more often want to know a fact from them or an opinion? Do you value what they can share because of their expertise, their experience, or their connection to you? Do you take pleasure just from thinking about them? When you think about them, how do you think of them?

The "date." Schedule a sexual encounter well in advance. In the

MANAGING SEXUAL FEELINGS

hours or days before, share with each other bits and pieces of what you plan to do. You might communicate that information through . . .

The back channel. Create new email accounts for each of you to use exclusively for intimate conversations. For some people, the back-channel will be a continuation of the kinds of things they say during intimate encounters. For others, it will be a place where they can finally express what they would never say out loud. Having this separate platform to go to can create great pleasure from anticipation.*

Fencing. For the first twenty minutes of an intimate session, "make out" with your partner, but make some unexpected, even deliciously frustrating exception, such as agreeing to no kissing on the mouth, no touching under the clothes, no speaking, no contact below the waist, or perhaps one or both of you wearing a blindfold.

The switch-up. Plan an intimate encounter with new ground rules: whoever is usually the initiator will play the more passive partner, and the usually passive partner will take the lead. Going against the norm can make sex feel mysterious and new again.

BUT IT FEELS SO GOOD

All our imperative-driven urges have this in common: they stand on constant lookout, hair-trigger sensitive for a call to action. But the desire for sexual stimulation is different. It comes calling not to rile us to some heavy task like defending ourselves or even to investigate curiosity or potential food. In the moment, it is simply a four-alarm invitation to feel good. The urge toward sex is the good-time buddy shouting in your ear to get out there and see what you can find. He comes out of desire dopamine, and he's sitting in your driveway, honking the horn, with the engine running. For this reason, out of all human desires, the urge for

---

* A word of warning: double-check the "To:" field before you press send. We've all sent emails to the wrong person. This would be an especially awkward occasion for that to happen. Better: Use a separate email platform or app for these messages and allow only one address in the account's contact list.

sex is the most likely to overwhelm reason, to discount what we want in the long run, and to blind us to risk and ruin.

The key to taming sex is preparing for that moment of choice and recognizing it when it comes. In that little bit of time between dopamine urge and dopamine action, between desire and control, it's too late to weigh what matters to us in the future versus the here and now—or to consider ethics, religion, risk of disease, uniqueness of opportunity, how attractive the other person is, how long it's been, and who's gonna know. To answer with anything but a guilt-driven "No!" or a hearty "Hell, yes!" we have to have considered the question long before. The answer, after all, is a consequential binary: Will we pursue this urge at this time or will we set it aside? If we pursue it, will we do so because we really want to, or because we are succumbing to a biological urge that swamped sense? If we reject it, how will we turn the regret and the urge into something of value? Frustration comes in lots of varieties, but sexual frustration is often the least manageable in its insistent demands for a consolation prize before it goes away.

Think of the dopaminergic challenge of sex as a knock at the door: we peep through the peephole to see what's out there, and we respond with a base reflex or a considered answer. The quality of that outcome, and our lives as they relate to sex, will depend on what we did to prepare for that consequential instant.

Chapter 9
HOW TO STAY TOGETHER

THE EVER-PRESENT WRECKING BALL

In the Western world, the legal aspects of marriage have long been less a necessity for daily life than an insurance policy for an especially stormy day that most couples will never experience. Until recently, marriage was just a legally codified version of cultural customs that had near universal acceptance, a public commitment to long-term monogamy and its benefits. The need to legislate it was primarily economic: formal marriage provides protection for the rest of the family when the breadwinning spouse leaves or passes away. When women were not allowed into the workforce, the state had a compelling interest in ensuring a husband could not financially abandon his wife and children without significant penalty.

That old model of "husband works, wife stays home" isn't nearly as common now, making financial protection much less important, but that's just half the purpose of the marriage contract. Our interest in the dopaminergic challenge of staying together begins with the other half, morality, and the fact that marriage and commitment are also less compelling now in terms of it.

Marriage unites anthropologist Helen Fisher's three components of love—lust, attraction, and attachment—under constructive constraint in a pleasant and public way. It is a contract, and the embrace of its blackletter law is happily announced in ceremonial language in a dedicated public event, then underlined with a celebration.* However, marriage's legitimization creates incentives less for the pairing itself than for *staying together*, which brings us to the point of this chapter. How do we stay together when we're dealing with dopamine's constant appeal toward the new and unknown, which are the elements that spark romance in the first place?

Those in a marriage are largely free to find attraction and attachment outside the relationship, but under penalty of law they may not act on lust. Why not focus the law on the other two? Because lust is the most powerful force of the three, an exercise in overwhelming biology-driven pleasure. Law, with its monopoly on violence, is relatively silent on simple attachment and attraction. But lust, tied as it is to hair-trigger biology, can quickly get out of control, and is sometimes only contained by another biological response, pain and the threat of pain. Mores change, but physical intimacy outside one's marriage remains a third rail, a Rubicon, a case of being allowed to look but not touch, as Bruce Springsteen put it. Nearly all marriages survive a little flirting with other people, but fewer remain whole after infidelity.

This is true even when the parties agree to "infidelity" beforehand. With the apparent rise of polyamory and open relationships in the twenty-first century, it's worth mentioning the original 1981 study of the challenge, Gay Talese's *Thy Neighbor's Wife*. Talese studied the many twentieth-century communities built on multipartner sexual relationships as well that those that devolved into such things, often

---

* Another force that matters here is the long-time (now largely vanished) taboo against cohabitation. While religious and moral objections were most commonly cited objections to such arrangements, the practical engine behind the taboo was that cohabitation without marriage provided no financial remedies for abandoned women and children.

by diktat of a (male) leader. Talese even participated in some of these groups. His observation: in every case, the sexually open communities quickly collapsed into dissolution, hostility, or violence. Which is not to say such a community isn't possible, but the overwhelming majority of evidence—per Talese, basically all the evidence—tells us that free access to sex eventually wrecks attachment, attraction, community, and family. Lust always finds a way. People who attempt nonmonogamous structures ought to be free to do so, but they are entitled to know what they are up against. Lust is a wrecking ball that blasts away at every level of a relationship. People in the West do not have a strong track record of success beyond the Western model of marriage.

However, now that the economic justification for marriage is, for many couples, completely gone, can love be enough? Probably not, but understanding the three elements of love may point us toward practical ways to stay together. Remaining in a relationship over the long haul seems to require a set of thoughtful practices and attitudes that reflect not an effort to defy our problematic tendencies but to recognize their power and redirect them.

THE CASE FOR A "ME FIRST" MARRIAGE

To be part of a successful couple, you must remain at the top of your own hierarchy.

Westerners have historically sought permanent pairing because we feel sexual desire for someone, plus intellectual and emotional curiosity, and then we feel something else: the desire to be a part of a couple. Perhaps it is the refined human preference for one-on-one comparison that draws us to things that come in twos, not threes or more. Once inside that relationship, we look for opportunities to practice self-denial—rejecting desire dopamine's call by, for instance, making a show of ignoring some attractive person in favor of doting on our spouse. We're hoping this makes our spouse feel our commitment via attraction and attachment, and reinforces their own. Self-denial is thus not only an element of maturity, it is a practical way to help a couple stay together.

As we try to convince the other person of our commitment, we often tell our partner, quite sincerely, that he or she is the most important part of our lives. But if we actually live that way, we're making a mistake.

In a literal sense, being self-centered in any relationship is not just good but necessary: if you don't take care of yourself, somebody else will have to, probably someone close to you. If you're not heavily committed to being responsible for yourself, you literally become someone else's problem.

In a relationship that lasts, taking responsibility for yourself is the only posture that holds things together in the long term. You "show up" to a relationship ready to give. That makes you a prime candidate for attachment. If the other person approaches coupling in the same way, you have two people who are aligned with each other and who are prepared to resist the attraction of distraction that dopamine makes us feel. By committing to address (and dispatch as needed) immediate, peripersonal demands as an independent being, we are freer to more fully engage with extrapersonal demands—demands that *a priori* require control dopamine's calculation, imagination, and commitment. Having prioritized self-responsibility we can move on to other things, and first on that other list ought to be the needs of the other person, whom we love. It is in this sense that they truly should be "first" among our priorities. We commit to looking after our own needs to the extent we are able, so we can give ourselves more completely to someone else.

Difficulties will arise because it's life and this is a marriage, but a foundation of "me first" in this important way means that when problems come, both people are well equipped to offer support and even unconditional caring. That security is one of the primary comforts of being a couple. The benefits, even joys, of marriage flow from this. For instance, if a couple decides to start a family, it's going to require individuals who are committed to showing up every day ready to give of themselves and be engaged with the other person. The "me first" approach is the starting point for addressing other couple challenges as they arise. It's also easy for children to understand and overlaps with most responsible approaches in religion and philosophy, if you're a parent who relies on those things.

People who are attached and in love will face complicated problems. Most of those challenges won't have easy answers. Some won't have answers at all, and we will end up living with the circumstances. But to give attachment the best chance of survival, we should cultivate it by looking after ourselves, the better to look after each other when required. That's vital, because when attachment goes, there's little reason to productively channel lust and attraction in any direction. "Me first" becomes "me only" and the marriage withers.

THE BEAUTIFUL LIE

Looking for a reason to stay in a relationship despite a history of difficulty? The passage of time can do more than heal the pain. It can rewrite memories in a more favorable light.

The lover in your arms is alluring, but nothing is as alluring as the lover you don't yet know. Dopamine lures us into love by giving us optimism for, even confidence in, things that are not certain. It makes us believe that a romantic partner could be the one we've always wanted, the one who can answer all our desires, the one who will finally make our dreams come true. When it comes to love, dopamine lays it on thick. If that weren't enough to contend with as we navigate romance, there's another neurological factor that can mislead us, not about the love that might be, but about the love we're already in.

Consider a computer's hard drive. The only time a document changes on the hard drive is when you choose to change it, then save it again. Your brain saves memories in much the same way: you bring a memory to mind, then savor it, share it, even recoil from it. But what happens next is probably not what you expect: after thinking about the memory, you replace the old "recording" in your brain with a new one, and this revised version has been altered by how you feel at the time. This is known as *reconsolidation*. Its purpose is to keep memories relevant to new information and experience, and to help us make sense of our past. Maybe you have warm feelings for your current romantic partner, but when you first met them you couldn't stand them. With those more generous, recent feelings in play, the memory of meeting them is going to be recalled, manipulated, and re-recorded based on that change. Perhaps that garish outfit and dumb hat now don't seem as garish and dumb. Maybe that crude remark they made to the waiter now sounds a little clever, come to think of it. Maybe that first date that seemed so awkward wasn't so awkward after all.

Except it was. The events of the date didn't change. What has changed is the recording of those events. Every time you think back on a memory, your brain is kneading the hard feelings (or possibly hardening soft feelings—it can work in either direction) into something that better comports with the way you feel now. Perhaps your feelings for your partner are bitter today, and that awful first date feels like "the good old days," softened in memory because you want to believe things weren't always this bad. Or you remember that night as worse than it was because you want to justify your current regret. Or the memory could go some other way.

Our search for real love with real people, grounded in real experiences, therefore starts with a couple of biologically internalized strikes against it—or in favor, if you don't mind your love seasoned with an extra-large dollop of illusion.

Yet our fuzzed-up-memory system is thus a catalyst for a working relationship, enabling us to diminish what we might not be able to forgive or forget. Over time, being unaware of the difference between what was and what we remember may be the thing that makes love possible at all. This way of living and loving relies on beautiful lies, or at least purposeful ones: we are drawn forward on dopamine-driven delusions of possible futures, and kept from falling away from relationships via perpetual, convenient improvement of our perception of the past.[79]

The abstract[80] of a related 2018 study begins by characterizing the eternal challenge of passion. In a subfield full of dispassionate description of the passionate life, the authors' opening is worthy of some kind of award for a below-freezing approach to affairs of the heart: "Romantic relationships are difficult to maintain novel and exciting for long periods of time," they write. "Individuals in love are known to engage in a variety of efforts to protect and maintain their romantic relationship." Indeed we are, friend. Indeed we are.

Having documented reality, the authors then suggest that the beautiful lie isn't just useful but is nearly indispensable. This may be revelatory to the public (though not its more cynical or realistic members), but to psychologists this is very old news. In 1988, Drs. Shelley Taylor and Jonathon Brown proposed that the truth isn't all it's cracked up to be, at least when it comes to mental health. While the prevailing thinking had been that a clear-eyed view of the world is a foundational requirement, Taylor and Brown argued that we see the world through rose-colored glasses. Not only is that all right, they assert, it's also important for the promotion of emotional well-being.[81] They called this false perception a "positive illusion." Forty years later, the authors of that 2018 study with the bloodless opening argued that the benefit of the positive illusion manifests itself in the form of successful romantic partnerships. The more often they observed positive illusion in members of a couple, the longer they stayed together, the more satisfaction they expressed in the relationship, and the less conflict they reported.

At the end of the Oscar-winning picture *Annie Hall*, Alvy Singer recalls the last time he saw his girlfriend, Annie, the one who got away. He asserts that we're willing to embrace a profound amount of self-delusion in order to experience love in any form, even when we're down to nothing but pure imagination:

> *It was great seeing Annie again, right? I realized what a terrific person she was and . . . and how much fun it was just knowing her and I . . . thought of that old joke, you know, this guy goes to a psychiatrist and says, "Doc, uh, my brother's crazy. He thinks he's a chicken." And the doctor says, "Well, why don't you turn him in?" And the guy says, "I would, but I need the eggs." Well, I guess that's pretty much how I feel about relationships. You know, they're totally irrational and crazy and absurd and . . . but, uh, I guess we keep goin' through it because most of us need the eggs.*[82]

THE COSTCO EFFECT

Everybody's heard the line "Parting is such sweet sorrow" from Shakespeare's *Romeo & Juliet*. It's the middle of a longer verse spoken by Juliet: "Good night, good night! Parting is such sweet sorrow / That I shall say good night till it be morrow." No phrase stays a part of a language for half a millennium unless it touches something at our core. Why do we find this line deeply romantic? We don't like goodbyes, especially with someone we love. And why "sweet"?

The answer is all around us: it's in the sample stands at Costco, the popularity of Victoria's Secret, and the allure of a stolen kiss. Sometimes only a little of something is enough to keep us hanging on for more.

Think about it: We want what we do not have, but how can that be? Our imagining needs a seed—enough of a taste of the distant thing to suggest that it might make us happy. Even a tiny sample can set us on that road. A half-bite of some new flavor of egg roll from the sample stand at Costco is enough to get lots of us to drop a five-pound box of them in our cart! We anticipate that the possession of the whole thing (or at least more of it) will be wonderful: since we have yet to experience it for very much or very long, we are free to idealize it.

What's more alluring, full nudity or a glimpse of skin under a negligee from Victoria's Secret? What makes you more mad with desire, being half of a couple alone with all the privacy you want, or stealing a passionate kiss in the kitchen at a neighbor's party? A taste of what's possible is powerful because this micro-bit of emotion can leverage the whole of the dopamine system into action. What's more, dopamine-driven anticipation is typically more satisfying than having anything in the here and now. To paraphrase Carly Simon, we can't know tomorrow, but that won't stop us from obsessing about it. Juliet had already experienced a bit of what was possible with Romeo. Each parting was a new opportunity for delicious imagining.

The power of anticipation explains the value of "separateness" in maintaining attachment. We often talk (and in Carly Simon's case, sing) about a committed relationship as "two becoming one," but those who have experienced such a bond know that it sounds better than it often works out. What we really want is a variation on that theme: two individuals entering into a union in which they also remain separate individuals. You're thus not replacing two people with one relationship, you're creating a relationship in addition to what's already there. The maintenance of separateness guarantees some distance and, therefore, some mystery, which gives us the thrill of dopaminergic lust and attraction. If there were no separate individuals, there would be nothing left to discover about the other, and with nothing left to discover, nothing remains to tantalize us toward *more*. In the words of Gregg Levoy, "It behooves us to stay hungry." We can do that if we cultivate the relationship in addition to, not instead of, cultivating selfhood.[83]

LET'S DO SOMETHING NEW!

We can reignite some of our dopaminergic passion by breaking out of routine and doing something new. That's why couples who find themselves in a rut often plan a romantic vacation. Usually "romantic" conjures up beaches, palm trees, and exotic cocktails in an oceanside bar as the sun sets in the distance. That's a great choice because we often daydream about remote tropical islands, and daydreaming is a dopamine-driven activity. But another option that's just as effective is a getaway in which you engage in activities chosen only because you've never done them before. They could be thrilling activities such as whitewater rafting or intellectual ones such as joining an archeological dig. They could also be simple: visiting a city you've never seen, learning a sport you've never played, or traveling in a way you've never tried, such as a train trip or cruise, camping, or trading hotels for Airbnbs.

You don't even have to go anywhere. You can engage with the new at the spur of the moment and at little to no cost. A few ideas:

- Buy a board game and learn to play it.
- Introduce yourself as a couple to the neighbors you've never met. Ask them over for dinner.

- Sign up for a night of learning painting, poetry, improv performance, or singing.
- Go to the mall. Each person gets a $10 bill. You have thirty minutes to buy something surprising for the other person.
- Go out to dinner, but order for the other person. Make it a rule that you can't order something you know they've had before. Whisper it to the server so it stays a secret until it arrives.
- Perform at a karaoke night. Especially if you're not good at singing.
- Pick a town, park, or destination about an hour away from where you live, which gives you time to anticipate together on the way there and review on the way back. If there doesn't seem to be anything there worth exploring, find something anyway.
- Visit a thrift store. Bonus points for drawing up a scavenger list before you get there.
- Read a book out loud to each other.
- Give yourselves fifteen minutes to write a love letter to the other person. When you're finished writing, swap letters and read them out loud.
- Cook something together.
- Throw a last-minute party. All you need is a few six-packs, some soft drinks, snacks, music, and the phone numbers for enough friends and neighbors to fill the space you have in mind.

The point is to put yourselves in a position where neither can say for certain how the other will react. Since you already like each other, the outcome will likely be a pleasure whether it validates your expectations or surprises you. You both get the dopamine-driven pleasure of wondering what will happen. At the end, you'll have been reminded that your partner can still surprise you in a good way. You'll also gain a refreshed, dopamine-driven anticipation for what other delightful surprises might be found in the other person.

The only caveat: You actually have to do something. None of this happens just sitting in the living room in front of your own screens. It's up to you—both of you.

PAM AND JIM'S CAMERA

Since dopamine directs our attention to the future, it can distract us from fully enjoying the pleasure in the present. It does this so subtly and firmly that we don't even know we've lost something. If it's true that all we have is the moment, then we should resolve to enjoy the moment and reinforce its value by doing something that highlights that value. On *The Office*, when Pam and Jim were getting married, one of them would stop every so often, point a pretend camera at the other person, and say, "Click." Their goal was to better savor this once-in-a-lifetime day, but it did more than that. It changed their priorities. They chose to stop, frequently, to take in the pleasure of the here and now.

We usually do a poor job of enjoying what's going on because we're thinking about what's next. Jim and Pam thought they were making memories, but they were doing more. They were ensuring that their experience of their wedding day would be filled with variety, depth, and pleasure, which, as a fortunate side effect, would make it more memorable anyway.

The here-and-now deserves your attention, and looking for moments to "photograph" brings you back relentlessly to the present. Taking pictures with an imaginary camera is a ritual that reminds you that whatever is happening right now is more important than what you're imagining in the future—because the future you're imagining may never arrive. That's why it's important to have the richest life experience in the moment, not just to enjoy recalling a memory later and not just to anticipate what might later come our way. Savor the experience. Let experience occupy your mind as it happens. Feel it.

Carry your imaginary camera with you wherever you go. Do it for yourself and your partner. Every day is filled with moments that are worth remembering—and also worth truly and completely experiencing.

HOW CAN YOU MEND A BROKEN HEART?

What about those times when love doesn't work out? While there are specific things you can do to feel better, the most encouraging thing to know is that your dopamine system will soon begin drawing your attention toward potential new partners. In the meantime, create comfort for yourself in the H&N system. Misery lives in the present but so does its remedy. Here are some things you can do to help clear those bad feelings, the better to get you back to an emotional state where you're more open to the dopaminergic pull of a possible new love.

Accept your sad feelings. A broken heart hurts. What you're feeling is normal. Your instinct is to push away the bad feelings. Don't. By allowing yourself to experience discomfort and even pain, you learn from the feelings and make the lessons a part of your outlook. From there you begin to move on from those feelings and the relationship itself, and become open to a new connection.

Acknowledge honestly how you felt about the other person, and perhaps still feel. You may be tempted to say that you weren't really in love, that you were kidding yourself that they were in love with you, or that the relationship wasn't valuable or even enjoyable. Those things may be true, but often they're only balm for a broken heart. Be honest with yourself. It's far more likely that the relationship had lots of positive things about it, but you're discounting them now because it's easier to lose something if it was worthless. Instead, think about the benefits you gained, what you were grateful for, the good times you had, and the lessons you learned for next time. Inventory those things—even write them down—and allow yourself to be grateful. Even if the relationship was costly, whatever you invested in feelings, time, and anything else is a sunk cost. It's gone and you're not going to recover it. Better to take the benefits that came from your time together than to bury the good with the bad.

Get some physical activity. When you do things like exercise, wash the car, or take a walk around the neighborhood, your body releases endorphin, a neurotransmitter that elevates mood. The last thing you may feel like doing is moving around. Do it anyway. Once you make the decision to get up and move around despite how you're feeling, it's a near certainty that you're going to feel better in just a few minutes. You'll be glad you pushed through the gloom. Doing it the first time will make it easier to do it the next time. In addition, consider planning a few physical activities each day. Not only is it good medicine to boost your outlook, it'll also improve your health in the long run.

Do something thoughtful for someone else. Heartbreak makes us focus on ourselves, but the remedy for heartbreak is not found solely inside. Look outward. Think of something you know a friend needs doing, then do it. Don't announce your intention and wait for permission. If you know it's needed, take the initiative. The more self-directed your help is for others, the more confidence you'll gain in overcoming your own pain.

Be your own cheerleader. When you sarcastically ask what your ex ever saw in you, answer the question sincerely instead. Write down the qualities that you find admirable in yourself—things about you that a future romantic partner might appreciate, or that a previous partner did. Maybe you always listened when they needed to talk. Maybe you cooked great meals, or showed up with unexpected gifts, or treated their parents nicely. Don't forget to consider subtle things, too, such as knowing when the other person needed "space," or being especially good at meshing their daily rhythms and habits with your own. You deserve credit, and at this point you're the only one who's going to remind you of your value in that relationship—what you did right. Claim it.

Chapter 10

PRODUCTIVITY AND GETTING AHEAD

HOW AMBITION BECOMES REALITY

Ambition without productivity is just a dream; productivity makes it a reality.

When we come up with efficient and original ways of doing things, we gain an advantage. This requires us to tame and channel dopamine—and it's worth doing. Even slightly improving a process can yield big benefits to the world, and those who figure it out usually end up with more money, prestige, power, and occasionally a place in history.

In 1855, the essayist Ralph Waldo Emerson confided this in his journal: "If a man . . . can make better chairs or knives, crucibles or church organs, than anybody else, you will find a broad hard-beaten road to his house, though it be in the woods." A generation later, just after his death, the idea was credited back to him in the form that would live on: "If a man can . . . make a better mousetrap than his neighbors . . . the world will make a beaten path to his door." Dopamine helps us recognize such opportunities, then motivates us to act and enables us to execute our ideas. It gives us the cunning and calculation to make something out of nothing or improve something we already have. Thus dopamine powers technological progress as well as our personal ability to "get ahead." It is the chief driver of productivity and improvement for individuals and for the world.

TWO KINDS OF THINKING . . .

Coming up with superior ways to get things done is a compound process. The first step is *divergent thinking*, the gathering of potentially useful ideas—key word, "potentially." Since the idea is to gather as many elements as we can in hopes of discovering something useful, the ideal approach is to take down the velvet rope and let almost anything in. If an element or idea isn't entirely crazy, we need to entertain it. At least, that's the goal. Most of the time, our brains automatically rule out a lot of things—because most of the time we need it to do so. If we paid attention to everything, we couldn't get through the day.

The next time you're walking down a busy sidewalk to get to a destination, pause to notice what you usually don't. You're not aware of the pavement. You're not paying attention to the faces of strangers. You're not aware of the patterns of the clouds in the sky, the sounds of traffic, or the feel of your clothing against your skin. Your brain takes note of all these things, but it doesn't elevate them to your attention because you don't need to know them. Your focus is on getting somewhere. Thinking about how your shoulders feel inside your jacket will distract you and slow you down, and your mind is focused not only on completing the task at hand but also on doing it efficiently.

Salience is the term we use to describe how important something is to us in a given situation. Our brains do this sorting—assigning levels of salience—without our conscious attention. When we're walking down the sidewalk, the traffic nearby is of low salience to us. But when we're crossing the street against the light, that roaring traffic is the most salient thing in our lives.

Divergent thinking is a dopaminergic process. Dopamine responds

PRODUCTIVITY AND GETTING AHEAD

to novelty, especially novelty that might make our lives better. It's always on the lookout for new resources. This is the realm of the desire circuit. Vigorous divergent thinking requires us to attach greater salience to otherwise ignored stimuli. We want to pay attention to the things we haven't noticed lately or noticed much at all, the better to collect more potentially useful things.

We want these things for the second step. Here we'll switch from gathering materials with little consideration for their usefulness to performing intellectual and emotional calculation. We'll try out these pieces in alternative functions and new combinations. We'll hold on to some elements, set aside others, and consider ways to connect them toward a goal. That sorting-and-solving is called *convergent thinking*, the complement to divergent thinking. We'll explore connections to see what kind of useful outcomes are possible, then we'll settle on the best one. Convergent thinking involves the control dopamine circuit that activates the prefrontal cortex (PFC), where we make decisions, solve problems, plan, organize, and carry out most other cognitive abilities. The PFC is logical, controlled, and systematic.

If I want to create a better cake, my divergent step will be to compose a list of ingredients that might be useful in a cake. Along with the basics like flour, eggs, and milk, I'll write out other things that go in some cakes and not others, like peanuts, pecans, cashews, pieces of fruit, and cinnamon. I'll also list things that aren't normal in cake at all, such as peanut butter, shredded radish, melted ice cream, black pepper, tomato soup, beer, and mashed potatoes. If I'm especially expansive in this step, I'll also include alternatives to the usual method of making a cake. Instead of baking it in an oven, I'll consider using a grill or a smoker, or maybe a way that doesn't require heating it up. Whatever I write down, the breadth of potential elements in the cake is limited only by my ability to overcome the low salience assigned to the many things that have faded into my mental wallpaper, the things I don't notice. Then, in the back half of the process, I'll focus my thinking, sorting through and combining some things and setting aside others. I'll test, experiment, and redo the run-through in my mind. From that calculating experimentation I'll come up with my new cake.

Nature uses this two-step process too. All kinds of genetic variability occur within organisms, but only those that increase an organism's fitness for survival get passed along to future generations, a process called *natural selection*, the result of which is constructive propagation of the species. Random genetic variability is analogous to divergent thinking, and the transmission of advantageous biological features (via "survival of the fittest") is the analog of convergent thinking.

Divergent and convergent thinking are powerfully complementary. As the poet Criss Jami said, "Create with the heart; build with the mind." The power in the two types of thinking is in their combination: divergent for open-minded gathering, convergent for highly focused calculation. Consider some people living with ADHD, which is associated with an overactive desire dopamine circuit, leading to impulsivity. As desire dopamine overperforms, the control dopamine circuit underperforms, causing difficulty with focus and attention. This is why people with ADHD present interesting conflicts in their abilities. For instance, some appear to have an advantage when it comes to founding new companies but are at a disadvantage when it comes to managing them—the former activity is more a divergent-thinking act while the latter calls for more convergent thinking. Overall, people with ADHD are much better at the loose, flexible forms of cognition needed for divergent thinking, but struggle with the more structured cognition required for convergent thinking—they tend to be better at coming up with ideas than developing practical plans for carrying them out.

Unlike with many other kinds of psychological processing, it's not important to cultivate a balance between these two parts. Still, the more accomplished you are at each of them, the better. We do our most impressive work using what writer Dorothy Parker called "a wild mind and a disciplined eye."

. . . AND HOW TO DO THEM MORE EFFECTIVELY

To be more productive, both personally and professionally, improve your ability to find and build original solutions to challenging problems.

To build your capacity for **divergent thinking,** which is the gathering of new ideas, try these techniques:

Brainstorming. It's an oldie but it's quite powerful. Pick the topic, problem, or concern at hand, then write down everything that comes to mind about it. Reject nothing. The idea is to give yourself access to ideas you might not ordinarily imagine. When I teach speechwriting, one of my topics is the original use of language. I ask students to craft unique metaphors for love. To move them into a more productive mindset, I first ask them to describe something far removed from love. I usually choose a chalkboard. When we get to the end of the exercise of calling out their descriptions, we have the material for some surprising metaphors. Some of the usual things I hear:

A chalkboard is easy to erase.
It comes in different colors.
You can fill it with whatever you like.
It's a platform for sharing ideas.
Everybody can see it.
If you abuse it, people will hear.
After a long time, it can wear out.

The metaphors are now easy to see and make for strong jumping-off points. *Love is like a chalkboard; what you put there can be easy to erase, easy to forget—but not always. Sometimes the marks are so deep and strong that they're never really gone, no matter how hard you wipe them away.*

Another aspect of the exercise can be a useful addition to your brainstorming arsenal. I tell my students that we're not going to limit ourselves to ten or twenty descriptions. We're going to take as much time as we need to literally fill the board. After we get through their first hundred suggestions, the students' well of ideas is exhausted. It's in these last moments that their descriptions move far away from the obvious. Instead of telling me how the board looks or feels, they start talking about obscure things that are typically not salient, such as the smell of the chalk dust near the board, and how the grain of the board varies on close examination. It is at this point that the students come up with their most original ideas.*

That happens for several reasons. The *incubation effect* is the powerful unconscious mind asserting itself, often as a sudden insight near the end of the process. While you've been grappling with the problem in your conscious mind, your unconscious mind has been working furiously with a larger set of data and more diverse possibilities. The phenomenon is similar to waking up with a good idea after your unconscious mind processed the problem during sleep.† Also, working for so long wears out our conscious imagination, and the power of *cognitive fatigue* kicks in, diminishing our ability to suppress unusual associations. In this case, not being able to block "silly" or "wrong" ideas is exactly what we need. Finally, the power of group dynamics is also in play. The combination of rapport and competition in a group can push us to reach for ideas that we might not otherwise find. It can also give us the confidence to share those odd ideas.

Brainstorming calls on dopamine in at least two ways. When we come up with a new idea, it can trigger a small dopamine release that in turn can create a feedback loop that drives the creation of more ideas. Brainstorming is also a focus on the novel and new, and finding such things triggers dopamine.

Reverse reasoning. Think about the opposite of what you're trying to achieve. Instead of asking how to solve a problem, ask how someone might *cause* the problem, or even how to make it worse. Then try to "solve" that "reverse" problem. The result: You will often see aspects of the original problem that were previously obscured. For instance, if you're trying to figure out how to make it easier for customers to return unwanted items, you might ask how you could make

---

* There's a story, likely apocryphal, that has long floated around Washington, DC, communications circles. A now-defunct website known for its catchy, clickbait headlines came up with them through calculated brainstorming. The rumor was that staff would sit in a room and write no fewer than two hundred headlines for every story. The one they published would always be one of the last ten or so they came up with—because that's when the truly unique ideas appeared.

† I'll expand on the connection between dreams and creativity in chapter 16.

returns more difficult. Walk through the process from the customer's perspective but in a new way. You'll be on the lookout for problems a customer encounters during a return—even minor issues because, after all, you're trying to create or amplify the trouble along the way. This can reveal new perspectives on and elements of the original problem.

One way to make returns more difficult might be to add a lengthy form to the front of the process. You might limit the hours and days when someone can return an item. You could make unreasonable requirements of the customer, such as maintaining the original packaging in perfect condition. You might add a restocking fee that makes a return more expensive than it's worth. Of course, the idea is not to explore the value of these proposals but to see how they point us toward pain points we might not have noticed otherwise—in this case, pointless paperwork, inconvenient hours of access, unrealistic rules on packaging, and unreasonable restocking fees. Now we have a first-attempt list of potential improvements. This approach has an advantage over traditional brainstorming because it leverages our interest in stories and narratives. That can naturally take us through a series of connected ideas in a process.

To build your capacity for **convergent thinking,** which is the synthesis of ideas toward a solution, try these techniques:

Assemble a list of pros and cons. One of the challenges of convergent thinking is comparing the quality of solutions. For each proposal, create a simple, two-column diagram, one for advantages and one for disadvantages. It's easy to do and can be a strong first step toward making a choice. Beyond giving you data on which to base your decision, it also guides you through quantifying benefits and drawbacks and prioritizing them. You can enhance the lists by categorizing each element as objective, meaning measurable ("It's red") or subjective, meaning the value may vary depending on opinion ("It's pretty"). Comparing lists for different solutions can quickly reveal gaps in what each solution can do and make clear which have benefits you cannot do without.

Create an Eisenhower Matrix. If your need for convergent thinking is less about the qualities of a solution and more about prioritizing actions—making a process or a schedule more efficient—consider this proven organizational approach.

Create a table with four categories:

1. Urgent and important (Do now)
2. Important but not urgent (Schedule)
3. Urgent but not important (Delegate or schedule soon)
4. Neither urgent nor important (Eliminate)

The idea here is that every task has a timeliness component and a value component: either urgent or not urgent, and either important or not important. In that case there are only four combinations to describe a task, which means we can assign every task to one of the four categories. Some things are obvious: if you have a broken leg, that's urgent and important and you should do something about it right away. If you have a report for your boss, but it's not due until next week, that's important but not urgent, and you should make a plan—a schedule—so you can have it done on time. In practice, this is a powerful way to establish workflow through a multistep process, work through a long to-do list, or create a schedule for each day across several weeks or months.

The problem with this approach comes when the urgent/important status of a task isn't clear-cut. Let's say you need to talk to your partner about scheduling travel for this weekend and it's already Wednesday. That's urgent and important, but it's not so urgent that you couldn't do it on Thursday instead. Or maybe you are miserable in your job and want to find a new place to work or even switch careers. That is terribly important and in some sense, urgent, but the world won't collapse if you delay a decision—which you know because you've already delayed doing anything about it for two months.

The key to success with the Eisenhower Matrix is in how you define the words "urgent" and "important." A good place to begin this is a thought experiment: define "urgent" as *attached to significant physical, financial, or emotional consequences if action is not taken within a few hours*. This would reserve the first category, urgent and important, only for short-timeline emergencies. Unless you drive an ambulance, that could be very limiting, because now most of the things you need to do are forced into category two: they are important and because of that, urgent by a different definition, so they need to be achieved in a few days or weeks. Perhaps we can modify our definition of "urgent" to be "requiring attention in the next five days"—or whatever timeframe works best for you. The point is that by attaching the requirement for action to a date certain, we gain a way to prioritize the things we need to do and then get them done.

As for categories three and four, they are a little easier to use and their definitions quite clarifying. If a task is urgent but not important, we can delegate it if we have the option to contact someone, or we can schedule it soon, calling on our revised "within five days" version of "urgent." That takes care of the third quadrant. As for the fourth, things that are neither urgent nor important can be eliminated. This is where a lot of your peace will come.

A few years ago, I was teaching a writing seminar when one of the students approached me afterward. She appreciated the material and wanted to gift me with some consulting on her approach to running a small business. "How long is your to-do list?" she said.

"Thirteen pages," I said.

"And how long have you had this list?"

"About two years," I said.

She told me to break it up according to the Eisenhower Matrix, adding that probably about half of it would end up in category 4. She was correct and I discarded those tasks. To this day I do not remember what they were, and I am better off for having abandoned them. After that she hooked me on the practice of scheduling my work on a calendar. Not only does it get done in a timely way, I get the dopamine-hit satisfaction of achieving goal after goal after goal, every day.

By building convergent and divergent thinking using well-defined methods, we can put originality in service to productivity—and we don't even have to feel particularly inspired to do so.

THE ART IN NUMBERS AND THE NUMBERS IN ART

The world is filled with successful, productive people who are good at both math and art. Michelangelo was a sculptor, painter, poet, and scientist. The son of a painter, nineteenth-century engineer James Nasmyth constructed heavy machine tools but also drew amazingly detailed lunar landscapes. Steve Jobs co-created the personal computer—the first business machine that was also aesthetically pleasing, even exciting. Former US secretary of state Condoleezza Rice is not only a diplomat but also an accomplished concert pianist. Comic actor Rowan Atkinson, the creator of the Mr. Bean character, holds a degree in electrical engineering. Actor-comedian Ken Jeong is a physician. Nagarjuna (full name Akkineni Nagarjuna Rao) is a prolific Indian actor and film producer who holds a master's degree in automobile engineering.

Some people who aren't good with numbers wear it like a crown. "I'm an artist," they say. "I'm not a math person at all, don't wanna be." Some number types proudly dismiss art. "I had to work hard to learn how to do this," they say, "but that painting/song/performance? Anybody could knock that out." Both are missing out. That their outputs are measured objectively or subjectively is, from a neuroscientific perspective, incidental. At the core, both call on the same dopaminergic skills: manipulating abstractions, engaging in divergent gathering and convergent reasoning, and mentally modeling what does not yet exist.

Familiarity with both art and math is becoming more important. Not only does simultaneous skill in both lead to greater opportunities, many modern pursuits now require it. Carnegie Mellon University recognized this when in 2022 they launched a new degree program for a Bachelor of Engineering Studies and Arts. To introduce the effort, the university showcased Professor Thomas Sullivan, who is affiliated with both the School of Engineering and the School of Music. His work includes sound processing for guitars that captures the individual strings (known as "hexaphonic processing") and the use of technology for including sound and light in art installations.[84]

If you're an artistic type who thinks you can't do numbers, give it another chance. Start with the basics of arithmetic. Pick up a book from the For Dummies series such as *Everyday Math* or *Math for Real Life*. If you already have a good grasp of arithmetic, you'll find that introductory books on algebra, calculus, or statistics are much easier to master than you expect—being motivated to learn these things for practical reasons instead of academic ones tends to make things easier. Also consider apps such as Duolingo—recently expanded to include math training—that "gamify" learning. You'll find that the step-by-step approach will leverage your dopaminergic reward system to sustain your interest. You'll also sense something familiar: The circuits you've relied upon in one realm will now fire toward mastery in another.

If you're a numbers- or reason-focused type mystified by the attraction of art, start with the thing you do best, logical thinking. Approach the "mystery" the way you approach any other problem, by inventorying its qualities (divergent thinking) and, from that exercise, finding ways to make sense of it intellectually (convergent thinking). Consider a painting. Don't ask questions that just lead to navel-gazing, such as whether it's pretty or what the artist was trying to say. Ask questions that leverage your skill for objective analysis. For instance, note how large or small the canvas is. Ask how this affects the level of detail that's possible to put in the space given the size of the subject. How deeply did the artist exploit the possibilities of the medium? What are the elements of the image? Is there a visual center to what you see, and if so, where is it located? What colors are in use? How do they relate to colors in real life? What is the effect of light? What is the texture of the painting and how does it affect what you see?

There are lots of reasons to learn to appreciate art, but one of the best is to enrich your appreciation of everything else. Art can give you fresh ways to consider what you already know and push you toward new thinking about the familiar. Regarding math, one of the great things about appreciating numbers and logic is that it allows you to recognize order in things that can seem chaotic. This in turn can reveal "methods to the madness" that enhance the abilities you already have. From either side, embracing the other realm will grow your productivity in everything, and draw you toward a richer, more fulfilling life.

THE MULTITASKING TRADE-OFF

We can do only one thing at a time. What we call multitasking isn't some workaround for this limit, it's just stopping an activity and moving to another. When you think about it that way, you understand why multitasking doesn't help you get things done faster. In fact, it slows you down, because the more transitions you make, the more time you waste. As much as 40% of the time you spend multitasking goes not to the work you're doing but to the act of switching between the jobs and mentally reorienting.[85] All this setting and resetting of our brains introduces error, slows execution, impairs learning, and makes us more susceptible to distraction.

But that's not the end of the story. Although multitasking won't make our productivity faster, it can make us more original.

Cognitive fixation is a bias toward the known. It occurs when we restrict our thinking to familiar methods and ideas—when control dopamine is firmly in charge, using the linear tools of rational thought to solve a problem and confining us to consider only what is, not what might be. Artificial intelligence researcher Tony McCaffrey suggests a tragic example:[86] The people on the *Titanic* overlooked the possibility that the iceberg could have been their lifeboat. Newspapers from the time estimated the size of the iceberg to be fifty to one hundred feet high and two hundred to four hundred feet long. *Titanic* was able to navigate for a while and could have pulled aside the iceberg. Many people could have climbed aboard it, McCaffrey postulates, to find flat places to stay out of the water for the four hours before help arrived. But the passengers were fixated on the fact that icebergs sink ships. Therefore they overlooked the possibilities of using the iceberg as a temporary "boat." They didn't consider its size and shape, or the fact that it would not sink.*

With an understanding of cognitive fixation, let's return to a consideration of multitasking. Switching from one task to another reduces

---

* For the sake of completeness, smart aleck, I acknowledge that not everyone agrees this could have worked.

the time we spend constructively working. But multitasking has a benefit: when we're yanked away from familiar thinking, we seem to open ourselves to fresh approaches. In one study, researchers directed a group of participants to alternate tasks on command. Another group was allowed to switch on their own. About half of the directed group successfully solved problems that required original thinking, compared to only 12% of those who switched at their own discretion.[87] Interrupting typical thinking—not just multitasking on our own but multitasking on external command—seems to push us toward unconventional problem solving. Sticking to the normal ways of thinking until we tire of them—multitasking when we've come to a dead end—doesn't have nearly as powerful an effect.

If you're working to solve a problem, try jarring yourself out of cognitive fixation. Set a timer and force yourself to switch between tasks every few minutes or, better yet, choose random intervals, or put the choice in the hands of someone else. What you lose in efficiency, you may make up in originality.

Chapter 11

TAMING SOCIAL MEDIA

JUST BEYOND THE BOTTOM OF THE SCREEN

It is impossible to imagine a purer demonstration—a purer *distillation*—of the addictive attraction of the dopamine chase than social media.

You're focused on your very own screen, scrolling through a literally endless stream of purposefully cropped pictures and blurbs crafted for maximum come-hither effect. Even better: The content you're reading (or, let's be honest, just glancing at) is being selected by an algorithm tuned and retuned in real time to promote things on the basis of what you most recently paused on. *But wait—there's more!* At any time you can indulge in the pleasure of offering your own hot take, which you can share with *the entire world*. You don't even have to convert your reaction into words. Simply click "Like" or post an emoji! You barely have to have an opinion at all. Just pause long enough to sort-of get what you're looking at, move the distal phalanx of your index finger, and tap. *Voilà!*

What keeps you looking and scrolling and reading is the ever-present, ever-updated, ever-promising promise that the next little blast of disposable intrigue—the blurb you can't yet see, can't quite imagine, the one you'll get a glimpse of in a second *if you just keep going*—there is every blessed possibility that the next one, still out of sight, will be even more interesting than the one you just saw, and be the ticket to more good feelings. In fact, that post *just beyond the bottom of the screen* might be downright fascinating! It might even be the thing you've wanted to know for *so long and now it's here and oh my gawd how cool is that?!*

Of course it won't be. It never is. The point is to grab your interest so hard that you give chase, not to actually provide you with anything certain. Then it's three in the morning and you're wide awake and you reflect for a moment on the return on this investment—in particular, on the two hours you just spent on Instagram with nothing to show for it. Then it occurs to you that you spend your time in this stupid way every day and if you were to add up all that time . . . and you decide that the total would be too depressing. So you put down the phone and try to go back to sleep, but you can't. So you pick up the phone again to see what fresh outrage some rando has whipped off for the outrage junkies on X . . .

Social media is the latest hijacker of the dopamine process. It is a Frankenstein's-monster combo of the feeling you get from mainlining drugs and the progress you make on a treadmill. In this application, dopamine is co-opted to sell advertising in the margins of your vision. One more time: dopamine's job is to ensure you're so enamored of the possibility of some better thing just over the rainbow—or beyond the bottom of the screen—that you will go after it, the better to charge advertisers for eyeballs and clicks. It's the first and only principle of social media marketing: if you're getting something for free, you're not the customer—you're the thing they're selling.

Most of us have a gut sense that social media can be bad for us, but it's only in recent years that we've begun to unravel the particulars. In a 2020 survey from Queens University in Canada,[88] researchers measured three elements of social media use: time spent, intensity of engagement, and "problematic" use, such as compulsive use, the inability to cut back, and negative impact on work or school. They found a correlation between problematic use and depression, and it mirrored the correlation with depression seen in people addicted to drugs. Given dopamine's key role in depression related to pleasure and reward, this suggests that dopamine may operate similarly in social media stimuli and drug addiction.

Pennsylvania senator John Fetterman has said in numerous interviews that social media made his depression worse, describing it on NBC's *Meet the Press* in December 2023 as an "accelerant, absolutely." Fetterman had been hospitalized for depression several months before, notably after years of typical social media interactions as well as hardcore "trolling," both giving and receiving. He told NBC he was most exercised about the personal negativity, which he said made him wonder if the abuse would follow him the rest of his life. "Look what it's done to me . . . and my family," he said. We all know the feeling of having been abused on social media or going too far ourselves. The bad feelings can stick around, and the consequences can go well beyond the platform.

In this chapter, I'll share some ways to tame the dopamine draw of social media so you can pull yourself away more easily from the grim delight of Instagram, Facebook, X, comment sections, and the rest. You can do it with just a few well-made choices that create a longer, more thoughtful pause between the desire dopamine blast and the control dopamine response. That gap is where consideration can live, and that important moment can belong entirely to you if you claim it. You can stop being an unpaid worker for influencers and advertisers making money off your personal data, lost hours, and despair.

OPT OUT OF THE INVITATION

Start by doing something that doesn't require willpower: use technology to eliminate the most attractive of the many valueless, time-wasting things that draw you to the platforms in the first place.

The most drastic option is the most effective: trade in your iPhone or Android for one without internet access. They still sell them, occasionally as retro-style flip phones but these days more often as a "children's phone" or as a device for an older adult who doesn't want or need browsers and apps.

Most of us, however, need the access to the world that smartphones provide, so we'll have to dig into the settings. By default, apps like Facebook and Instagram sound an alert when a new post comes in. That's an explicit signal for your dopamine system: you *will* feel the urge to investigate it, and you will never acquire the ability to consistently ignore it. For those moments when you are away from your phone (and it's rarely more than moments, right?) or for when you don't hear the alarm, the app icon features a little circle with the number of unread messages. Turn all this off. On an iPhone or an Android phone, go to Settings, then Notifications. Scroll down to each social media app and turn off Sounds, then turn off Badges. Below that, you may also find a Show Previews option. Turn that off to eliminate notices on the lock screen. With this in place, you'll go to a social media app because you decided to yourself, not because the platform rang a bell to make you come running.

LOCK YOURSELF OUT

Comedian John Mulaney tells a funny story about a grim subject: trying to beat his addiction to cocaine. Mulaney figured he'd cut himself off from being able to buy the drug by making it impossible to get cash. He called his accountant with instructions that unless his email request for cash was copied to his doctor, the accountant should ignore him. Mulaney immediately regretted doing this, but it worked anyway: he knew he could call his accountant and rescind the order, but he was too embarrassed to do it.

When it comes to social media access, you can do something like that with your phone and browser. Built-in apps such as Screen Time on iPhone can be used to limit the times of day and total amount of time you can spend on social media applications. Yes, you can override the limits just by returning to the Settings panel and turning them off, but there's a way around that. Ask someone you trust to change the master password for your phone, then give it the "Mulaney move": tell them to forbid you to change the social media settings. There are also third-party apps that allow you to lock yourself out of another app according to a schedule, then lock yourself out of changing that schedule for a while or forever. Depending on the app, you can defeat the system only by switching to another browser (easy), asking someone else to rescind your request (a hassle and maybe embarrassing), reinstalling the application or your entire operating system (risky and time consuming), or deleting the social media account and starting a new one (for many, unthinkable).

TURN DOWN THE "JUST FOR YOU" FACTOR

Next, lessen the dopaminergic attraction of your social media feeds.

You have the most influence over your social media feed because you choose the friends, celebrities, influencers, products, and companies you follow. But platforms also refine your feed based on what you pause on and how long you stay there. The system that incorporates this feedback is called the algorithm. It's typically talked about in hushed terms, but it's nothing more than an elaborate database application[*] that, in practice, matches the tags of things you already like with the tags of potential new content. The algorithm is mysterious only in the sense that users don't know how it weighs your preferences, but even if we did know, the problem would still exist. What we need to do to combat this is turn off as much feedback to the algorithm as possible.

Under Settings, usually in the category of Privacy or Security, select Tracking and turn it off. You may also find a list of apps there that need

---

[*] Though by the time you're reading this, algorithms will also use artificial intelligence to forecast things you likely want to see.

resetting to "deny permission," or you may need to open each app and check the settings there.* An internet search for instruction on this may help. Defaults change too frequently and subtly to document them step by step in this book, so it's worth digging into each social media app to limit what it shares. The less you allow the system to harvest about you and your habits, the less it will be able to tailor material to your tastes, which will make it a little easier to tear yourself away from the feed.

CAUSE SOME CLIMATE CHANGE

We all know we spend too much time on social media. But why do we do it? It's more than just wanting information, entertainment, or connection—we can get those things elsewhere. The real culprit is dopamine. Social media is designed to trigger us with a constant stream of surprises and novelty. Every like, comment, or new post is a potential "win," making social media into a slot machine that keeps us coming back for more.

Think about it: you scroll through your feed, expecting to see the usual updates from friends and family. But then, boom! You stumble upon a hilarious meme, a shocking news story, or a juicy piece of gossip. That unexpected reward hits your brain with a surge of dopamine, making you crave more.

Of course, there's nothing inherently wrong with amusement. Social media can be a fun way to connect with others and unwind. But when it starts to dominate our lives and distract us from more important things, it's time to take control.

One way to break free is to identify and eliminate the sources of empty stimulation. Those accounts that rely on outrage, negativity, or misleading headlines to grab your attention? Unfollow them. Look for

---

* On an iPhone, you can go further by selecting Apple Advertising under Privacy & Security and turning off Personalized Ads. At the time of this writing, this seems to affect only Apple-served ads on their own applications, but that could change at any time.

patterns in the content that leave you feeling empty or frustrated, and reclaim your attention.

For instance, start recognizing that clickbait headlines are a lie. When you see one, tell yourself that. You might even say it out loud: *That's a lie. That's a trick. That's not news.* In the same way we condition ourselves to not respond every time we hear someone shout, "Hey!" in a public place, we can change the way we respond to clickbait, first as a conscious choice and ultimately as habit. Think about it: There's nothing on Instagram that you absolutely need to know right now or can't get somewhere else, no significant threat or gain that depends on a click-through, and nothing so imminent that it is presented to you only on a social media post.

By eliminating your feed's biggest polluters, you change your social media climate and improve your life. How to recognize them? Here are some model lines that are a dead giveaway.

- "This little-known fact could change your life today . . ."
- "Here's something you didn't know . . ."
- "This will shock you . . ."
- "Watch as this . . ."
- "Hear what FAMOUS PERSON said about . . ."
- "You won't believe what is about to happen to your retirement account/investment/savings/job . . ."
- "5 (OR 7 OR 10) Things You Can Do Right Now to . . ."
- "FAMOUS PERSON Reveals Secret to . . ."
- "Here's why you are having a PARTICULAR BAD EXPERIENCE . . ."

One more thing: Get rid of all the news and opinion sites in your feed. Really.

Your first instinct will be to get rid of only "polarizing" pages, but that won't be enough. Even when the goal of a news organization really is to inform you, that aspiration will always be in competition with their need to get your attention—to earn clicks on social media and generate ad revenue. They inevitably do this with provocative headlines written to incite your interest in (a) seeing an opponent embarrassed or (b) having your own opinion validated. Get rid of them all. There are other places to read the news besides social media. Try visiting the news organization's website directly. While you'll find many of the same clickbait headlines, you now have an advantage: you weren't lured there by a dopamine-stimulating headline "randomly" presented in your stream.

As you tame the attraction of social media, you might find using it to keep up with the news a reasonable carve-out. Despite the allure of those headlines, you're getting something valuable in exchange, right? But the price is much higher than you think.

THE JOY OF HATE

It's not just depressing political news that makes me want to put a fist through my screen, or at least your screen . . . It's also the open sewer that is social media. I'm not "on" Twitter, as in, I don't post. But I search it plenty, the way a lion might show up at a savanna watering hole to spy its next meal. Except instead of asking if I should consume an impala or a wildebeest, I ask myself if I should consume a migrant-vs-native European fistfight, or a supermarket checkout line beatdown, or a girl-on-girl school hallway tussle . . . I lose all faith in humanity after 30 seconds of watching. So imagine how despondent I feel 30 minutes later, when I'm still watching . . . We are surrounded by a-holes, it seems. Made all the more apparent if there's a mirror in the vicinity, since the a-holes are often us, craving more of the violence and nihilistic exhibitionism that is desensitizing us all.
—Matt Labash, "Be This Guy, Instead of the Angry Jerk You're Becoming"[89]

It's easy to forget that interacting with each other via social media is a very new thing. The original social network, Facebook, didn't become accessible to the general public until 2006. In the short time since, change has been profound, and one change that may seem interesting but minor turns out to have been the first domino in a long collapse.

Until social media, it was nearly impossible to have a significant relationship without meeting someone in person. Now we have friends, clients, employers—and opponents—we've never been in the same room with.

This has consequences.

If you were an employed adult in the 1990s or before, you knew that some meetings were too important to have over the phone. You'd fly across the country or around the world so you could sit at a table across from someone else. There was nothing you could not have shared as a disembodied voice over the telephone, yet on some occasions you went to considerable time, trouble, and expense so you could talk sitting just a few feet apart. Why do this? Because a trip to meet someone in person not only lets us use the considerable power of nonverbal cues to make our case, but also demonstrates to the other person that we consider them worth the effort to get there.

Romantic partners used to see each other in person at least once in a while, or there wasn't a relationship. It's an old joke in the US that "my girlfriend who lives in Canada" is cover for not having a girlfriend at all. (It's the subject of an entire subplot in the hit musical *Avenue Q*, which features the song, "My Girlfriend, Who Lives in Canada.") As for platonic friends, keeping up the relationship required purposeful decisions to write a letter, share a phone call, or meet somewhere, which is why long-distance friendships tended to fade away. In those days, friends came and went more frequently than they do now. Friends tended to fit into a season of your life. As time passed, you stopped going to the trouble of staying connected. It wasn't that you didn't like the person anymore, it was that the work it took to stay in touch was no longer worth the benefit that came from having them around.

But what has changed most consequentially is how we treat strangers. For most of history, we gave others a wide berth for privacy. Most Boomers can name people they've known since childhood whose political affiliations remain a mystery. Need directions to the grocery? Ask away. Wanna know who they're gonna vote for? Not so fast. Yet today, via social media, we don't have to ask those kinds of questions because we regularly broadcast the answers. And when we're not broadcasting, we're inquiring—from a distance, and that is key. We ask intimate questions of people we've never met and criticize them for answers we don't like. It's now scarily common practice to identify someone who's expressed some heterodox opinion, gather a mob, and punish them with public reproach or economic isolation. How did we get here?

To meet someone, you first have to have something in common. Opportunities for such an introduction in the physical world tend to come from and point toward a positive connection, such as working in the same office or striking up a conversation at a bar or a show. But on social media, opportunities often come in situations that invite disagreement. Consider Facebook, where you might come across a page for admirers or opponents of a political party, candidate, ideology, or social issue. If you go there to find a friend, you can. But some people show up to fight, which is far more stimulating—and safer to do online than it ever was in the real world. Being at a distance removes the incentive we once had to engage sincerely, exercise caution, and offer grace. The physical-world culture of human interaction came with guardrails that the online world doesn't have. Being nasty in person can earn you a sock in the jaw. Being nasty anonymously and at a distance comes with no consequences at all—at least, not immediately.

Online conflict (a term that makes it sound nicer than it is) is attractive. Agreement is stasis, but conflict comes with the opportunity for a series of dopamine hits. Without the disincentives to physical and emotional violence that are built into real-life contact, we are free to be mean and destructive, then meaner and more destructive. Short of hunting someone down, there won't be physical comeuppance because there cannot be. In this way we are no longer incentivized to be patient in the face of offense. When you're hiding behind a screen name on a social media platform or in the comment section of a website, the only limits on what you might say are what the platform or your conscience impose on you.

This is how social media so often enlists the dopaminergic capacity for imagination into the joy of hate.

On social media, people are abstractions, and dopamine is what allows experimentation with the abstract. With people as abstractions, you can be as extreme as you wish to be. Humans with feelings become indistinguishable from bloodless ideas. We start to think of an online opponent not as a person or even an avatar representing a person, but as a target that just happens to react. Except we're not sword-fighting with disposable intellections. We're tearing down other people as unlikely to disengage as we are. This becomes a race to the bottom of discourse and dignity.

The nature of the dopamine system is a perfect match for the nature of the platform, and the result is a supercharged pursuit not of truth but of *more*. People chasing the thrill of the putdown no longer encounter the guardrails of the physical world that would hold them back in person. Why is culture so coarse just now? When people don't matter, culture doesn't, either—or civility, norms, customs, or rules. So we turn our guns on whatever irritates us in the moment, sometimes (and purportedly) in the name of justice, but to some extent just to get a dopamine hit by getting a rise out of people who disagree.

Yet, you ask, aren't some of us online sincerely? Oh, yes! And crazily outnumbered. For every person thoughtful in this way there are tens of thousands of others tossing grenades for the fun of watching them explode. When's the last time you encountered someone on social media who was reconsidering their opinion? The worst of our internet-driven instincts make us callous to human dignity, to the hard-won worth of an ordered society, and to reasonable avenues by which we can improve each other's lives. The distance of the internet has made us lesser, colder citizens. We demand more of each other but expect less of ourselves.

Our at-a-distance encounters with other people add perpetual novelty to the mix—anything could happen, because people are capable of anything. One has to ask where the benefit lies in what we're doing. We're arguing with strangers and probing friends for personal opinions that don't change anything for the better if they don't go our way. More often than not, the replies divide us, and the difference of opinion likely didn't matter to either party until they knew it existed. When you find yourself tempted to post or reply to a comment online, ask yourself: Is this exchange about a meeting of the minds or a chase for dopamine-hit conflict? Consider how rarely you've seen anyone come away from an online debate with a changed mind, or even a more informed perspective. It's more likely one or both people lost track of any noble goal early on. That's not because we're bad people. It's because we're humans.

To anyone with an interest in policy, politics, or debate, social media and comment sections can be irresistible. But think before you post. If you're there to change minds, know that there's little record of constructive engagement. The lack of a physical-world connection makes persuasion or even learning that much less likely. The sincere pursuit of knowledge quickly falls into competition with the desire to defeat the other person—more specifically, to use whatever tools are handy to make them give up, including illogic and personal attacks. That's how strong the attraction is to the dopamine buzz. Most online discussions aren't discussions at all. They are pursuits to catch a neurotransmitter thrill: cast your opinion or insult out there and stand by for what comes back—and sometimes it's stimulating validation! Other times it's pushback or insult, but even that prompts another round of dopamine thrills. Things elevate quickly because out there in the ether of the internet, where the natural limits of human contact are not a part of the environment, no one has much reason to do anything but escalate, escalate, escalate. As Vonnegut put it, "So it goes."

There's no casual engagement to be had safely, no indulgence that doesn't damage yourself and others. In this corner of social media, the only winning move is not to play.

Chapter 12
OBSESSING OVER THE NEWS

LEARNING FROM "A YEAR WITH NO NEWS"

Dopamine pulls us relentlessly toward the new and unfamiliar. You can learn everything about your spouse, become an expert in your professional field, read all the books by your favorite author, and learn all the songs from your favorite band, but eventually you'll run out of new stuff to know. One thing is assured to show up in your awareness everyday with something you didn't expect: the news. This is a large part of why highly dopaminergic people are attracted to politics and policy: not only do they provide the constant dopaminergic stimulation of figuring out a problem, they also deliver a never-ending stream of fresh factors and faces to be figured into those solutions, some new data and permutation for control dopamine to chew on. That this new news is available constantly with just a swipe down an iPhone screen makes it that much more appealing to both desire and control circuits.

Maybe you've seen the moment from the series *Parks and Recreation* where the intense Ron Swanson, played by Nick Offerman, places the most memorable food order in sitcom history. Ron, a man who profoundly loves steak, has just learned that his favorite steakhouse is closed. Bereft, he ends up in a diner only to be served a thin slice of something barely approximating meat. "This isn't a steak," he says to the counterman. "Why would you call it that on your menu?" He then places his famous order.

"Just give me all the bacon and eggs you have," he says. A half-second later he calls back the server. "I'm worried what you heard was *give me a lot of bacon and eggs*. What I said was *give me all the bacon and eggs you have*. Do you understand?"

In 2017, I cut myself off from the news. I'm worried that what you just read was "In 2017, Mike watched a lot less news." What I wrote, and meant, was "I cut myself off from the news." All of it. I blocked news sites from my browser and phone. When news came on the radio, I switched stations immediately—not after the headlines, but right away. I refused to pick up a newspaper. I stopped reading op-eds and analysis, which was quite a feat for me. (I live in Washington, DC, and came here to work in politics. I don't do that work anymore.) I teach courses in public relations, persuasion, professional writing, speechwriting, and the crafting of op-eds. I've written speeches and editorials for political figures including candidates for president of the United States. Keeping up with the news was not only something I had come to do instinctively, it was also something I took a lot of pleasure from. I could count the ways: I liked being knowledgeable about current events, having an informed opinion, being qualified to seriously debate issues, and being able to present the other side of an issue in a way an opponent would recognize (more dopaminergic pleasures, you'll note).

Still, I felt that politics and policy took up too much of my internal life, so I decided the best and fastest way to find out and maybe fix it was to go "cold turkey." It was also the most difficult, and I knew it would be. I imagined that walking away from the news would be like cutting myself off from food or family.

I was wrong.

As the end of the year rolled around, I realized that the world had somehow continued to turn without my careful attention. There remained enough interested human beings across the spectrum of opinion that my departure had not led to any decline in social progress or civil discourse. I saw that all that reading and worrying and intense opinion-making I used to do had barely affected it. And unlike nearly every other news obsessive, I had spent years doing more than reading and opining. I'd worked on Capitol Hill and all over K Street* in the thick of the fight for and against various laws, regulations, rulings, and appointments. And now I wondered, *Why?*

This had not been the miserable divorce I expected. I thought I would miss keeping up with the news. I did not. What I thought I was gaining, knowledge and the pleasure of analysis, I came to see not as the purpose of my news reading but as an incidental byproduct. My pursuit of those things had been just another dopamine chase all along.

Keeping up with the news had let me enjoy a range of dopamine-driven delights: the mystery of what the next headline might bring, the allure of what else I might learn and the intellectual battle that might ensue, and the anticipation of what my new analysis might be, an analysis that would not in fact enrich discourse for the public or among my friends (who, I realized, didn't care much in the first place). But the emotional and intellectual energy I was spending on all this—the time, as well—was a profound investment with no payoff, for the world or for me.

Long before my year of no news, well after my work became far less focused on current events, my keeping up with the news had become nothing more than a hobby—but I hadn't figured that out. No one was getting any value from it, not even me. What I thought of as civic responsibility and self-improvement was a well-disguised version of a crossword puzzle. I was having fun while deluding myself it was something more.

The sour cherry on top: this pursuit had strained friendships, a few to the breaking point. I hadn't changed any minds. I had only reinforced the positions of people who agreed with me and made those

---

* "K Street," referring to the location of many lobbyists' offices, is Washington, DC, shorthand for the influence industry.

who disagreed feel more alienated. I wasn't chasing some noble good. I was just getting pulled into the black hole of chasing approval on social media. All I had ever done was check the number of likes, and that wasn't making anybody's life better, especially mine. Looking back, I wonder how I would have spent my time without social media to urge me on. I was just another thoughtless runner on the high-tech treadmill, mistaking my short-term interest for a lifetime achievement.

Here's where the experience gets especially interesting. If I had been pursuing the news toward some single issue or cause, my dopamine chase would have been an efficient and satisfying engine to keep me on top of the game. But it wasn't. It was just an amusement—and while there's nothing wrong with an occasional dopamine chase as amusement, it was nothing more. Aristotle taught that activities not aimed toward some greater, valuable end leave us empty. Psychiatrist Viktor Frankl created an entire school of therapy around this idea, as you'll recall from chapter five. They were both right.* I'm glad I figured this out before any more time had passed.

WANNA TRY IT YOURSELF?

Consider spending your own year without news. It's hard to begin but easy to sustain, much easier than you expect. Simply do what I did: get rid of access to news on your devices, then decide that you will walk away when you encounter it "in the wild." Before you begin, choose two or three other activities that you will turn to when you're tempted to check the news, such as playing a word game you like, calling a friend, or reading a book. One more thing: Keep this campaign to yourself. Otherwise your friends will be self-conscious about mentioning the news in front of you, and that will just push your desire back to the top of your mind. They're likely to mention things from the news. That's fine. Acknowledge it and change the topic. If the topic doesn't change, excuse yourself and leave.

---

* We'll take this on at the end of the book.

OBSESSING OVER THE NEWS

You may be afraid that you're going to miss something of life-or-death importance, but you won't. It's impossible to completely insulate yourself from the news. If it's all that important, someone will put it in front of you. Also, your conception of "life-or-death importance" is going to change. We argue a lot about who the president is, which party controls the House and Senate, and what various and sundry legislation will mean for jobs and prices. These things affect our lives, obviously, but the effects are usually subtle, slow, long term, and only one cluster of factors in a stew of many. On most matters the impact is so unnoticeable that the only reason we know they're in play is because . . . we read a headline on our phone.

If you need proof, set a date on your calendar three months from the beginning of your effort. This will be the day you allow yourself to read some news, assuming you still want to. But I can tell you already what you will find on that day: the headlines will pick up close to where they left off. The drama went on, change was minimal, and the effect on your life was close to zero. Yet the effect of being away from the news for only a few months will have given you a new perspective on what matters, on how much more control you can have over your urges to "keep up," and on what truly qualifies as a matter of life and death. Many of us who think we're chasing a better world are, it turns out, chasing something else.

Chapter 13
ONLINE PORNOGRAPHY

THE MAGNITUDE OF THE PROBLEM

Online pornography is as powerful a force for personal damage as culture has ever seen. It answers an immutable human urge with nothing productive, yet it is available for free, it can be consumed in solitude for hours at a time, and it offers intense, good feelings with little biological downregulation or emotional pushback. Online porn is a slick machine powered by dopamine-driven attraction and it has no natural governors to tame its effect on you.

That means it can create spectacular trouble in your life.

Online porn can exercise a nearly irresistible pull. Saying no to it is more than a matter of willpower, preference, or moral choice-making. Some people can make the craving a manageable part of life. Far more cannot. The experience of problematic pornography use (PPU) has a lot in common with what we see in people with addictions to alcohol and drugs. It is wrapped up with biological proclivities in addition to dopamine attraction, such as arousal via norepinephrine, the downregulation of pleasure via endorphin, and the attachment aspects of oxytocin and vasopressin, which can create a sense of bonding with pornography itself. In some cases, PPU may also be a reaction to anxiety, depression, social isolation, or trauma such as childhood abuse.[90]

Online porn is a problem for far more men than women. Women suffer from PPU, but it is men who more often have a problem. About 28% of men report problematic use of pornography[91] compared to only 3% of women[92]—though this number may be artificially low due to the greater cultural stigma assigned to women over the use of porn.[93]

Online porn leads to sexual dysfunction. In a 2021 study of 3,400 men aged eighteen to thirty-five, 20% had some degree of erectile dysfunction. The more a man used online pornography, the more likely he was to have ED. More than a quarter of these men found a way to deal with it, but here's where the effects of downregulation make another appearance: the way many restored their previous quality of erection was by watching more porn, more extreme scenes, or both. This approach works for only a while though. As with alcohol and drug addiction, you may be able to feel better by drinking more, but pretty soon you're drinking all the time just to keep some feeling of normalcy, and then even that fails. Women may suffer decreased libido and arousal. This is likely driven by the same mechanism in men, with overexposure to pornography leading to desensitization. Women may also find it more difficult to achieve orgasm, probably due in part to the difference between the expectations set by pornography versus the reality of partnered sex.[94]

It damages relationships. The American Enterprise Institute's Survey Center for American Life reports a correlation between pornography use and decreased satisfaction with one's sex life.[95] *Science* magazine notes its association with an increased rate of divorce.[96]

The younger you start, the worse it will be. The accessibility of porn at an early age seems to have a negative influence as well. The earlier in life a male begins masturbating to porn, the more likely he will be to have ED. Among those who started masturbating to porn prior to age ten, nearly 60% of them experienced ED.[97]

ONLINE PORNOGRAPHY

The problem is more prevalent than you think. Twelve percent of all websites contain pornographic content,[98] with 69% of men and 40% of women reporting they encounter it annually.[99] The internet has become the largest distribution network for on-demand sexual stimulation, with annual global revenue estimates ranging from $1 billion[100] to nearly $100 billion[101]—five times the revenue of the NFL.[102] More people struggle with the problematic use of pornography[103] than gluten intolerance,[104] and the likelihood of experiencing PPU as a life-altering problem is higher than the risk of dying from any of the top ten leading causes of death* in the United States.[105]

Yet PPU is more often treated as a moral failing rather than a health concern, and individuals often hesitate to confide in friends or healthcare professionals. It remains a taboo subject, and this silence creates a significant barrier to seeking help.

The internet has made possible the most significant sexual opportunity in history: Easy access to free pornography decouples a basic human experience, sexual stimulation, from the need for other people. That said, this decoupling began long before the internet.

THE TWO GREAT DECOUPLINGS

The arrival of online pornography seems like the greatest point of inflection in sexual history, but it's not.

On June 23, 1960, the FDA approved for sale the first oral contraceptive for women. To say this transformed the cultural landscape is to engage in wild understatement. The "Pill" disconnected reproduction from full-contact sexual activity. Today we take this for granted, but in 1960, the chance at no-condom sex that didn't come with a ticket to the baby lottery? It was the stuff of a brave, new world. The Pill made intimacy spontaneous in a way it had never been before. We slipped the surly bonds of awkward precoital inquiry in favor of uninterrupted

---

* Heart disease, cancer, COVID-19, unintentional injuries, stroke, respiratory disease, Alzheimer's disease, liver disease, diabetes, and kidney disease.

getting on with it, anytime. Sex, which in all its risky portent had loomed large over every romantic relationship, could now be reduced to just another physical act.

An epoch spanning the whole of human history came to a close. We no longer needed much caution toward sex and its consequences. In neuroscientific terms, desire dopamine could now tell control dopamine to go pound sand.

Ethical, religious, and moral considerations (plus the threat of disease) still existed but their importance was no longer reinforced by the rigid biology of making babies. We used to have to weigh the temptation of dopamine-promised physical pleasure right now against the burden its result might impose on the rest of a life. Not anymore. Call it the first great decoupling. Online porn is the second.

Before that online revolution could occur, the ground had been seeded in the 1980s, the decade that gave birth to the videotape rental industry. What was previously available only in seedy movie theaters was now playing on your TV. Forget "girly" magazines. This was live action! This brand-new (and dopamine-boosting) option made masturbation more appealing, more intense, and more satisfying. Pornographic videos didn't make masturbation as intense as sex with another person, but it narrowed the gap enough that for many, especially those who might be socially awkward, sex alone could be a reasonable substitute for sex with someone else, and it was available anytime. The only human contact required was picking out a tape with other people in the room and handing money to a clerk.

In the late 1990s, the practicality and affordability (and later the necessity) of home internet service swept away even those obstacles. Since online porn is financed mostly by advertising, much of it is free to consumers. Home internet service would quickly become ubiquitous. And the internet meant you no longer had to interact with anyone at all. This brought real parity to the competition between masturbation and intercourse. Thus, online porn really is the second great decoupling: the disconnection of intense sexual pleasure from the need for even a casual human relationship.

The problem with online porn is therefore a matter of the desire dopamine urge toward a foundational human desire unmoored from external constraint. Never has something existed that was at once so desirable and so free from immediate negative consequences. It really is the perfect beast, built not on human connection but human disconnection.

HOW TO REGAIN CONTROL

The battle to win control over a problem with online porn is best carried out on multiple fronts. Here are several tools—weapons—to give you an advantage, but don't pick just one. You'll defeat the urge more often by combining them. The most successful effort begins with creating a significant barrier between you and online porn, buttressed with other tools behind it. Online porn is a problem rooted in interactive technology, so several of the things I've recommended in the chapter on social media will apply here too.

Lean on technology. Almost no one succeeds at this on willpower alone. Don't assume you're going to be the one who actually guts his or her way to a solution. Instead, plan for the temptation before it arrives: make porn more difficult to access.

The quickest way to get a little breathing room between you and temptation is to use technology. Unlike social media, which we usually access via app, porn is most often accessed through a browser. That means you'll need to install a blocker that is sensitive to content or that maintains a list of sites and DNS addresses that it won't let you reach. Some blockers even block particular pages within a site. Others also let you specify what kind of content you'd like to block. Given that new blockers appear regularly and old ones change names or go away, I'll forgo listing names here and advise you to just do a search for "porn blocker."

It's always possible to work around a blocker, but putting one in place on your devices will at least add an extra delay between you and the porn you want to see. What you do with that time is crucial. Recognize it for what it is, a moment between stimulation and action that provides the "observing self" with opportunity. Use that time to think about what you want in that moment compared to the fulfillment you seek in the long run for yourself.

But noble aspirations are rarely enough to get us where we want to go, so don't stop there.

Be ready with something else to do. The attraction of pornography is hard enough to beat even with a distraction. But with no distraction at all, it's impossible. In the dopamine-driven battle between *what I want right now* and *what I want in the long run*, giving up pleasure in the moment loses nearly every time. So plan ahead.

Make a list of things you can do instead. Don't make it a list of alternative websites to visit. Don't substitute a computer game or a news website for porn. Plan something that forces you away from web technology. It doesn't have to be an intense activity or complicated, it just needs to take you out of the situation for a few moments in a way that feels satisfying or even rewarding.

Some kind of physical activity is always a powerful tool. Choose something with a purpose and that consumes your attention. Have a goal, such as doing a predetermined number of stretches, pushups, or chin-ups. Even a walk around the block might help. In addition to the ad hoc use of physical activity, consider a daily regimen of exercise. A scheduled routine of even a few minutes has salutary effects in lots of areas that can impact our desire for online porn and our ability to resist. People who exercise reduce their stress levels, which can lower the need to use porn and masturbation as a release. And you may actually replace the compulsion for online porn with an urge to exercise, an action-and-reward mechanism in the brain that is much more under your direction, and comes with public, measurable benefits.

Confide in someone as an accountability partner. We talk with our friends and family about a lot of things, many of them personal, intimate, and embarrassing. But it is still rare that we share with other people the attraction we may feel for pornography. If you have someone you can trust with this, tell them what you're struggling with and ask them if they'd help you by allowing you to call when you're dealing with the impulse. This will be awkward. Do it anyway. Since you're reading this book, you're probably the kind of person who would be honored to help someone through a struggle like this. Your friends probably feel the same way—after all, they're *your* friends. If you don't want an accountability partner, or you truly can't find one, consider just calling someone to chat when the urge strikes. Better yet, do a video chat, since that will prevent you from looking at porn while you're talking.

Track your triggers and learn from them. Compulsive feelings, panic attacks, and other stressful moments often seem to come out of nowhere. If you can identify a trigger, you can make it easier to regain some control.

Each time you feel the urge to look at online porn, ask yourself what happened to you in the previous few minutes, then write it down with a time and date. Did you just return from a certain event or place that may, somehow, trigger an urge? Did you just feel a particular stress or interact with someone specific? Were you busy? Were you bored? What were you thinking about? A person? A problem? An occurrence? A relationship? Don't dismiss anything as too trivial or unlikely—remember, this is the information-gathering phase of the effort, not the analysis. Write down the intensity of the urge too. Use a scale from 1 to 5, with 1 being mild and 5 being the inability to resist. In a few days or weeks you may find that many of these moments came right after the same kind of action, thought, or event. With that knowledge, you can prepare yourself to react in another way or avoid that trigger as much as you can.

Some influential experiences are buried a little deeper. The problem with porn may have roots in your attitude about the opposite sex (or the same sex) or some childhood trauma; in those cases, taking a deep dive with a therapist into this concern can help. If the thought of addressing things in this way is discouraging, recall what we know from ACT. You don't always have to delve into a traumatic memory or a problem to find a way to feel better. We can often acquire techniques that will help us function well without unpacking painful experiences.

Even if you don't find a clear cause, you can still benefit. In a 2020 study, a group of men kept a diary of their pornography viewing with the goal of reducing their use of it. At the end of the testing period, a majority of the men reported less negative thinking, improved self-acceptance, and fewer feelings of guilt and shame. The researchers concluded that the mere act of keeping track of what they were doing forced the participants to think more seriously about their porn use than they had before. This self-reflection improved their ability to deal with online porn.[106]

At the very least, get out of the place where the trigger appeared. Walk to a convenience store and buy a snack or a bottle of water—a few calories from a bag of chips is a better choice than indulging the attraction to porn, especially early in your effort. Get away from the place it happened for a bit and distract yourself. Carry a blank notebook and draw a picture of what you see. Grab a musical instrument, take it to a park, and learn some chords or practice a song. A ukulele is easy to keep around. A harmonica fits in your pocket. Or keep the instrument in your car so you have to spend a few moments walking out to get it. Whatever you choose, leave your phone behind. Every time you distance yourself from the physical location where you are drawn to online porn, you make it easier to pull away the next time.

Consider "rebooting." One of the most common approaches to treating addiction is abstinence or "rebooting," meaning simply stopping. It's difficult, but it works for many people for alcohol and drugs, and it appears to work for online porn too. This doesn't necessarily mean "white-knuckling" it—it's not a pure exercise in Nietzschean will to power. And you can use any method you like to cut yourself off; how you do it isn't as important as the fact that you're doing it at all. When you shut off online porn completely, the results can be profound.

In a 2021 study of previous efforts at rebooting, psychologists at Nottingham Trent University in the UK found promising results for those willing and able to take on such a tall order.[107] Staying away from online porn for as little as two weeks was associated with greater commitment in relationships and an improved ability to recognize and analyze personal compulsions. The researchers also cited several clinical reports in which participants reported a significant decline in sexual dysfunction such as low sexual desire during partnered sex, erectile dysfunction, and difficulty achieving orgasm during partnered sex.

Remember the big takeaway. If you decide you have a problem and want to take action, don't just try to cut back. Resolve to stop, ultimately, whether you get there over time or all at once. Don't try to do this alone because that's a needlessly hard path—and a lonely one too. Instead, assemble an arsenal of tools that support each other, and thus support you.

Chapter 14

BUYING STUFF

CAN SHOPPING BE AN ADDICTION?

Shopping, even just window shopping, provides coping and comfort. It delivers a direct shot of dopamine over and over as you imagine things you might—and that's the key, of course, *might*—come to possess by pulling out the credit card and completing the transaction. You get the delight of visualizing what your life might—*might*—be like if you transformed these things from wished-for to real: designer clothes, sophisticated electronics, impressive cars, high-powered computers, slick cellphones, glamorous jewelry. The restorative power of shopping is so powerful that we call it retail *therapy*. And you don't have to go to the Hermès store or the Tiffany counter to get the buzz. For some of us, a walk through the treasures at Goodwill can do the trick, or loading up a virtual cart at Amazon.com in anticipation of that delicious moment when you might, *might*, press the yellow bar that says, "Place your order."*

---

* The author is not immune. In early 2024 I did this with the latest iPhone on the website of my cellular provider. I put it in the "cart," then "abandoned" it so many days in a row that I eventually received a personal call from a Verizon agent named

There's a debate among professionals over whether certain unwanted behaviors satisfy the technical definition of addiction. It's an interesting discussion if you're not the one who's suffering. The rest of us shouldn't concern ourselves. Most of us who suffer with such behaviors in our lives are just trying to get better. We are troubled by some feeling, urge, or compulsion that we want to be rid of, and we'd like to know if it's something we ought to seek help with or if we can carry out some effective improvement ourselves. In other words, are we dealing with addiction or a simple need for more self-control?

Some of us shop more than we should. We're not spending ourselves into oblivion, but we occasionally buy more than we really need, wanted to, or can afford. If you feel bothered by your behavior, that's reason enough to change it, but it's wise to recognize when seemingly addictive attractions are best addressed professionally. Consider these questions:

1. Do you spend many of your waking hours doing this behavior, planning for it, arranging your life around it, or recovering from its effects?
2. Do you rely on this behavior to feel normal?
3. Does it cause you mental or physical harm, or put you at risk? If so, do you keep doing it anyway?
4. Have you tried and failed several times to cut back?
5. Do you sacrifice time with work, family, or other personal interests to make room for this behavior?
6. Do you hide this behavior or its frequency from others?
7. When you stop, do you feel sad or irritable?[108]

The more you answer "yes," the more likely your unwanted behavior is a kind of addiction, and professional help may be the best way to go. If your answers are mostly "no," it's likely you can make a change without professional support. Either way, if you feel your attraction is

---

"Marissa." She offered to complete the transaction and be my "personal Verizon representative going forward." (I bought the phone.)

slipping beyond your control, reach out to someone who can direct therapy and prescribe medication. Therapies such as CBT and ACT, often combined with antidepressants or anti-anxiety drugs, can reduce the frequency, impact, and attraction of the feelings you're having. There are other methods too. Proven approaches include mindfulness and distress tolerance via dialectical behavior therapy, using interpersonal therapy to find ways to connect with others, and traditional family therapy. Support groups make a difference for many, and can be all they need. For others, the problem may be tied to financial issues, and some counseling on money matters, probably accompanied by a few lifestyle changes, can fix things. Whatever you choose, there's no need to suffer, because treatment in this area has a strong record of success.

If you'd like to take on this dopamine-related problem on your own, how should you begin? Let's start by exploring how shopping came to such prominence in modern life, and consider it beyond its negatives. Then let's explore the psychological nature of the problem. Finally, let's consider particular techniques you can learn and practice to help yourself.

SPENDING OUR WAY TO A BETTER LIFE

Compulsive shopping, also known as oniomania, is encouraged by the times. Media is underwritten by advertising, which surrounds us with slick appeals to buy things. Technology further boosts this urge as much by customized advertising to our interests as by making possible products that didn't exist only a few years ago. Not only can we lust after stuff we can hold in our hands, we can also buy things we'll never even touch, such as digitally delivered services from video-streaming sites and music providers. The twenty-first-century market is a seller's dream and a buyer's paradise.

Many of us are tempted to condemn it all as the triumph of crass consumerism, and that may be what it is. But like so much of what we've examined in the realm of dopamine-driven challenges, what seems to be entirely negative—here, the temptation of over-shopping and "too much choice"—comes with considerable benefits too. Just as dopamine leads us unto temptation, it has also led us unto spectacular productivity, yielding unprecedented prosperity and opportunity.

In 1984, before we started carrying around efficiency-boosting cellphones and laptops, US median real income was about $59,000 per household. By 2023, that figure had increased by 37% to $80,610,[109] the number of hours worked weekly was lower,[110] and the size of the American economy had more than tripled.[111] Even life expectancy went up.[112] Until the late twentieth century, consumers had a choice of only one telephone service provider. Now there are over a hundred—and cell phones we can carry everywhere have replaced landlines tethered to the wall. There were once only three commercial broadcast networks, and they went off the air around one in the morning. Now there are hundreds of streamers with on-demand programming 24 hours a day. Until the Carter administration opened the airline travel marketplace to choice in 1978, jetting off to anywhere was truly affordable only to the wealthy. Now anyone can buy a ticket to anywhere on the cheap.

The burden of choice is significant, but we'd have less rich lives without it. Increased economic activity has led to improvements in many aspects of our quality of life, even longevity. Day-to-day consumerism may feel inelegant, and extreme consumerism comes with real problems, but the record shows that, overall, choice beats the alternative.

MY BRAIN MADE ME BUY IT

People who shop more than they can afford to, or who rely on it for a satisfaction that never comes, are experiencing the classic dopamine promise in the retail space: if you go get this thing, you will be happier than you are right now. There are other factors in play, of course. For instance, people who turn to shopping for comfort sometimes have low self-esteem and are more easily influenced than others.[113] That could naturally promote certain kinds of purchases. If your mood depends on the opinions of the people around you, buying things that those people like or promote can signal that you're on their side. But those motivations are secondary because we know that in terms of brain chemistry, it's the dopamine hit that matters. In a 2007 paper published in *World Psychiatry*, the official journal of the World Psychiatric Association, Dr. Donald Black identified "four distinct phases" of compulsive buying disorder.[114] The first two, anticipation and preparation, describe the activities of the desire circuit and the control circuit in the classic order they operate. The last two, shopping and spending, describe consummatory and acquisitional experiences in the here and now.

After the buying, though, there's not always great pleasure taken in the having. Compulsive or problem shoppers often buy things they don't need or use. Sometimes the purchase never comes out of the shopping bag. Instead, the shopper spends more and more time buying and browsing. Shopping may take on ritualistic qualities in terms of when and where the shopper goes, what the shopper does before, during, and after, and even what the shopper carries as a habit, a compulsion, or a "good-luck charm." The cycle of purchase and letdown plays out again and again: a ride up toward pleasure, then back down to disappointment and guilt.

Sometimes we engage in a behavior because it gives us relief from some mental intrusion called *obsession*: a thought, urge, or image that is "recurrent and persistent," that we don't want, and that is marked by "anxiety or distress."[115] People with obsessions often relieve them with actions called *compulsions*: repetitive responses such as handwashing, making a gesture, or saying some word or phrase over and over.

We know from the study of obsessive-compulsive disorder that compulsions may arise from dopamine activity in two areas: the *dorsomedial striatum*, which is related to goal-oriented learning, and the *dorsolateral striatum*, which is related to habit.[116] Here's how this seems to work. Goal-oriented behavior happens in response to the incentive of a reward, like skipping dessert in order to lose weight. Habitual behavior is an unthinking response that doesn't require a reward, like saying, "Bless you" when someone sneezes. If these two systems get wrapped up together, a behavior that used to require a reward could become a habit—a compulsion. When I scratch my nose, the itch goes away. But if scratching my nose became detached from the relief it used to give me—the reward—I might nonetheless keep scratching my nose for no reason, doing it as a habit. What used to be a matter of cause and effect would now be "just because." To fix such a problem, we have to find ways to wear down the habit.

RETREATING FROM RETAIL THERAPY

Fortunately, there are effective techniques to deal with shopping obsession.

Like the online gamblers in the study who still wanted to carry their phones, you have to do something other than cut out all temptation, because temptation is built into the landscape of living. Minimize exposure at the start, but to achieve long-term success, start changing your reactions to temptation. Don't ask yourself only how to avoid encountering the problem. Focus on how you aspire to behave—on how to be the person you want to be. Ultimately you will reduce how much shopping you do and how often you do it. You can diminish the attraction by wearing down the habit and understanding its source. The result really can restore your shopping to a choice instead of a compulsion.

Put your credit cards out of reach. Stop carrying your credit cards. Keep them at home. Take them out only when you're going to purchase something you've decided in advance that you want and need. And remove the ability to pay with mobile devices such as your watch or phone. Extreme tip: Make it even harder to impulsively use your credit card by putting it in a plastic cup filled with water and keeping it in the freezer. You'll have to wait for it to thaw before you can use it,* and that will give you time not only to reconsider the purchase, but also to reflect on the effort you're making to cut back. At worst, you'll try to rush the process along by chipping the card out of the ice. My advice?

---

* I know what you're thinking: I'll microwave it. Nope. You don't want to microwave your credit card. You'll short out the chip and melt the card. There's also a tiny antenna in it made of metal that will self-destruct by way of a spectacular little light show. Yikes.

If you try to bust it out that way, do it in front of a mirror—and video yourself. It'll give you an arresting image of what you're willing to do to indulge your impulse. You won't like how you look, and that may give you the motivation to choose otherwise in the future.

Plan alternatives for your free time. Most of us don't have to think too much about what to do with a free hour. The problem with that is we often don't think at all and instead fall thoughtlessly into habit—which is what we are often trying to avoid. Once again, a remedy for previous challenges is a potential remedy here too. Before the time comes to choose how to spend your time, prepare a list of those choices. Categorize it by length of time. A free half hour? Walk the dog around. Do a set of stretches. Call a friend. A free ninety minutes? Go to the gym. Walk to the library. A free afternoon? Pick a distant location for any of those activities and drive there to do them. Stick your list to the refrigerator, because we are more influenced by something we experience with our senses than by things we have to remember, especially when we're pushing back against an unwanted impulse. In other words, if it's on paper instead of just in your mind, it's already more "real."

Don't shop alone. If you're making a run to buy something, bring along some insurance against excessive shopping in the form of someone who knows what you're up against, who doesn't have the problem themselves. They can be the out-loud version of that little voice in your head that's encouraging you to resist, and you'll have incentive to do so because you want to show them that you're doing well at your effort.

Write down why you want to stop. The next time the urge to shop strikes, sit down with a pencil and paper.* Write down the numbers one through ten along the edge of the page. Next to each, write a single-line reason why you intend to go shopping. You may find it difficult past the first couple entries. Keep at it. By forcing yourself to think for a while, you dive past the obvious answers to the deeper and more important reasons. And if you can't find them, that in itself is an answer. As with CBT, the idea is to find the causes that drive the effect,

---

* Don't do it on a keyboard. Writing by hand slows us down, and that gives us more time to think.

meaning the reasons that you continue to shop even though the things you buy don't bring you satisfaction. By limiting yourself to a single line for each instead of, say, a paragraph, you make yourself cut to the heart of the matter. You won't be able to avoid saying what you mean by hiding behind a fancy sentence. This kind of self-examination, undertaken in the moment that the attraction arrives, can help you build resistance by pushing against the habit with action and reason.

A little understanding can go a long way toward realigning your motivation. It's not easy, but calling on various approaches, such as joining forces with others and creating time between urge and action to call on the "observing self," can give us additional advantages in this effort. We're always going to have to buy things, so complete avoidance won't fix this problem. But repeated success at resistance makes it easier to resist each time after.

Chapter 15

GAMING

I DON'T NEED A LIFE, I'M A GAMER

More people play video games[117] than go to the movies[118] or watch live sports.[119] In terms of revenue, the video game industry is bigger than those two businesses combined.[120] It even swamps pornography.[121] The driving force behind all this mental and financial investment is more than just "because it's fun." A survey by the Entertainment Software Association[122] makes it clear: gamers play to reduce stress, anxiety, and feelings of isolation. They say gaming makes them feel happier and helps them "navigate difficult times."

Gaming is a uniquely intense anticipatory experience, holding our attention by what we can call a "dopamine drip." Unlike activities where a single challenge leads to a single reward, video gaming is a series of encounters that are endlessly varied and always new in at least some small way. These repeated handoffs from desire dopamine to control dopamine come at random intervals, each transition promising the possibility of at least a small reward, and sometimes something much bigger. This pathway pulls a gamer along to the next step, over and over, for as long as they want to stick around—or feel compelled to stay.

It is a sophisticated version of a slot machine: every bet comes not with a promise but with the far more alluring feeling of "maybe." There's more: the game adjusts the experience to better appeal to the player based on what things get the biggest response. As artificial intelligence improves, this on-the-fly customization will get more precise and more effective. Along with that, complex games once available only on gaming consoles are more and more often accessible on cell phones. Add the use of advertising to make games available at no direct cost to the player, "loot boxes"* to entice a player to spend money for extras, and the constantly advancing technology that embeds this "dopamine drip" into an ever more immersive sensory experience, and it's obvious why computer gaming is such a force.

The critical difference between getting hooked on video games versus, say, exercise, is that exercise is just a thing you do, not an ever-evolving experience that interacts with you. Exercise has a natural end point; you decide when you're done, unlike some video games designed to keep you playing indefinitely.

USING YOUR SKILLS TO WIN A WHOLE NEW GAME

To tame problematic gaming, we can start by calling on what we have learned about other dopamine-driven issues, including romance:

- Like romantic attraction, the attraction of gaming is fueled by novelty, escalation, uncertainty, and anticipation in pursuit of reward.
- When we are looking for the right person for a relationship, and when we manage sexual feelings, we often have to sacrifice what we want in the moment for what we'd prefer in

---

* Some games give players opportunities to claim a "box" of valuables, usually things like weapons, armor, and character outfits. To find out what's in the box you've won, you have to pay for a "key." Maybe you'll get something great and maybe you won't. That's the point. It's a dopamine tease. For subscription-based games, it's a source of additional revenue. For free games, it supplements or replaces ad profits.

the long term. Gamers who are trying to create new habits make a similar calculation.
- When we're trying to maintain a relationship, we must sometimes cultivate alternative pathways to satisfaction. When gamers try to replace some of their gaming with other activities, they have to look for pleasure in things they may not have done before.
- Gamers, like partners in long-term relationships, have to learn to recognize when they're falling for a "beautiful lie" about the experiences they're chasing or the ones they've already had.
- As with obsessive use of social media and problematic pornography use, problematic gaming is overcome in part by exploiting technology for our own benefit instead of letting it continue to exploit our human nature.

One more bit of encouragement: The thing that makes someone a good gamer can also help that person overcome a problem with gaming. Strong gamers are fluent in pattern recognition and mental management. Reclaiming control over gaming calls on the same skills, only this time in service to wearing down the compulsion or habit.

To help you reclaim what you may be losing to video gaming, consider three approaches: getting support from other people, better managing your time, and better understanding yourself.

LEVEL UP YOUR SOCIAL LIFE

Use a human connection to moderate your gaming habits.

Get some company. Most gamers say that 75 to 100% of the time, they play games physically alone, meaning they play a single-player title.[123] At the same time, many gamers play because they like having other gamers as friends. I spoke with Drake Greer, a former graduate student of mine who for a time was a gamer playing competitively.[124] "I have met plenty of my best friends through video games," he said. "They live in other time zones and I've never met them in person.

Sometimes we're not even playing, it's just the catalyst to hang out and talk." This, he notes, presents an opportunity to enjoy gaming more while spending less time doing it.

Resolve that some portion of your gaming time must be spent playing with other people in multiplayer games, or in the same physical room. Find somebody for an hour or two a week—don't set an impossible standard. And don't be afraid to tell them why you're doing it. They may feel the same way. Just the presence of other gamers can help regulate the time you spend in the activity. "Once one person logs off," said Drake, "it can create a domino effect. Sometimes that realization, *It's just me now*, is enough for me to move on from a game that's become a habit. It's helped me realize it isn't as fun."

The other person in the room doesn't even have to be playing. "I'm starting to see a social media trend with millennials and Gen Z: the gamer boyfriend with the bookworm girlfriend," said Drake. "It's still bonding but in separate activities."

Find a friend to support you. In the same way an accountability partner can help you maintain your goals in cutting off the use of online porn, an accountability partner can provide similar support "to game or not to game." The person doesn't have to be another gamer. Depending on your situation, someone who knows nothing about gaming might be even better. They'll rely on you to help them understand what you're doing and why stopping "cold" in the middle of a game comes with its own problems. This will give you another opportunity to think through the challenge you're facing and the choices you have to address it.

HIT THE PAUSE BUTTON

Your time has value, and you deserve to be in charge of it. Here are ways to take control of the time you spend gaming.

Figure out how much time to spend on gaming. Start by tallying up how much time you spend on it now. The PS5, Xbox, and Nintendo Switch can do this for you. At the time of this writing, you can access the information via the home screen, your profile, or parental controls. Steam provides an even more detailed breakdown.

This is an easy way to track your time, but there's a better solution. (It's also the only solution, other than an app, if you play on a PC or phone.) When you game, write down the start and stop time. After three or four days, total the hours per day you've spent gaming. Using this as a baseline, choose a slightly lower number as your first per-day goal. Don't be extreme. A time diet is a lot like a calorie diet: you can't cut radically or all at once and expect to succeed. Start slow.

Let's say you game two to four hours a day. That's an average of three hours, which is 180 minutes a day. Fifteen percent of that is twenty-seven minutes, roughly a half hour. Set your new limit as two and a half hours. Each week, lower your gaming a little more until you reach your goal, whatever you decide it should be. Drake Greer suggests setting two alarms, a "soft stop" and a "hard stop." The soft stop, ten to twenty-five minutes before the hard stop, is a signal to finish any outstanding missions or objectives, or to help you decide if you have time to fit in "one more level." That's important, since many games don't let you pause or quit without penalty. It's easier to stick to the hard stop if it doesn't cost you the hard work you've put in up to that moment. It also gives you greater control over your actions.

Choose the time constraints that are right for you in light of the other goals you have for your life. You'll find plenty of guidance on the internet about what the "average" gamer does, but remember: you're not average. What's right for you is right for you because you're you. If you decide your ultimate goal is a limit of an hour a day, three times a week, so be it. If you decide you can afford two hours a day every day—well, if that fits in with your other priorities, it's the right choice for you. Your goal is to regain control over gaming so you can gain greater control of your life, reject unwanted impulses more successfully, and through those things, live a more satisfying life.

Schedule your gaming to fit your priorities. You don't have to follow a hardcore schedule in order to cut back. Be creative about finding time to reclaim. Limit yourself to evenings or afternoons, or whatever time of day suits your life. Plan your day's "intake" of gaming the same as you would with calories but remind yourself that your goal is only a goal, not a command. It's okay be flexible. For extra support, block access at certain times with parental controls or third-party apps.

"Even though I may still game in small increments—on a plane, now and then on a lazy day—vacations or travel days are big resets," Drake said. "I've noticed that my gaming isn't as frequent due to having that habit broken up by other activities. What is usually a daily habit for an hour or two, I'll go a few days without. Then, returning back to where I live, I find myself more selective about which gaming activities I choose."

Anticipate those times when the urge is particularly strong. Look for patterns in your original time tally. Do you spend more time gaming in the evening? In the morning? After work? On the weekend? In response to stressful conversations? After meals? Knowing when you are most likely to want to game will help you direct your attention elsewhere before you're most vulnerable. You'll meet your goal many days, but probably not all of them. You're human. Don't let falling short become the end of the effort. Forgive yourself, get back up on the horse, and try again.

Cut whole days out of your gaming schedule and notice how that affects your resolve. During those "off" times, direct that energy toward something else that matters to you. "Over spring break of 2024, I was out of town and had the opportunity to game on a lazy Saturday," said Drake. "Instead, I decided to spend the day reading an entire book, something I can't remember doing in years. I even spoke with my grandparents. The ability to game was still there, but since I was in a different setting and would have had to switch platforms to my laptop or my Nintendo Switch, I found it easier to push away from gaming."

HACK YOUR HABITS

Consider what else might appeal to you in addition to gaming, and think about what you find attractive about gaming in the first place.

Plan something else to do before you need something else to do. When you say no to gaming, be ready to say yes to something else. Have an alternative at the ready, just as you would if you were working against the attraction of social media or online porn. What to do? Say this out loud: "*I'd rather spend less time on gaming and more time on . . .*" and listen to what comes out.

As we've observed with other dopamine-related challenges, physical activity occupies us more thoroughly than a purely mental diversion, so climb off the couch (or out of that fancy gaming chair) and get moving. Pick up the house. Run the vacuum. Walk the dog. Practice your tennis serve. Take a walk. Go to the batting cages. Hit the driving range.

Another thought from gamer Drake Greer: "Most often I game at night after coming home from work or school. It's when most of my friends are online. However, right next to my door, I have a whiteboard listing alternative tasks. It's really easy to walk in feeling tired and just switch on the TV. But this whiteboard reminds me: *Hey, these are some of the things I have to do.* Are they all appealing activities? Not always (do dishes, start laundry) but some are fun things I could be doing (organize books, finish writing homework or reports, read the unfinished book I have on the side table). Seeing the list is a reminder to counterbalance the BIG reminder of my PS5 and TV sitting in my living room. This helps me choose differently."

Transform gaming from a default time filler into a special event. If you can move your system to a room that you don't need for other things, you create a feeling that gaming is a more unique, more special, more worthy-of-respect activity than the daily, tossed-off time filler it has become. Choose a room in your home that is the only place where gaming is allowed. This can help you segregate the activity in your mind from other things. But don't choose the bedroom. The temptation to go beyond your time limits will be strong if you're climbing in bed and the opportunity to wind down with a game is at hand.*

Rehearse the moment of "temptation" before it arrives.

---

* If you game on a phone, this approach won't work. Consider using the techniques that work for social media issues.

You have a list of alternative things to do other than gaming. This ensures you don't have to stop and think of something in the moment when the attraction is strongest. Now rehearse making that choice. Practice saying yes to the different thing and no to gaming. By doing this you get to experience the feeling of self-denial and thoughtful consideration—being the "observing self"—which is one of the skills in ACT. This equips you to act on your reflections on the feeling instead of the feeling itself. It also gives you time to remind yourself of your larger goal: reclaiming control over your time and productivity.

Think of when the "moment of choice" is most likely to come, where you will be, and what you might be doing before it arrives. Don't confine this to an internal dialogue. Act out the experience and talk to yourself out loud. The sensory stimulation during practice makes a stronger memory, which will make you more effective when the time comes to act. (And when that time comes, say it out loud then too.)

Think about why you game—and why you game so much. Gamers are pretty united in why they game: it's a uniquely powerful activity for reducing stress and raising mood. But that's a different matter from why you might keep gaming after you've felt the relief you sought. What makes you keep going? Ask yourself straightforward questions and write down everything that comes to mind. Don't filter yourself. Treat it like brainstorming. These questions might include:

- How do you feel when you're gaming?
- How do you feel when you win?
- How do you feel when you lose?
- What feels different about gaming versus other things you do, such as playing or watching sports?
- Why do you choose gaming over other things?
- What amount of gaming would you consider "healthy"? Why?
- When you feel you can't stop when you want to, what does that feel like? Be specific.

GAMING

Reflecting on these questions and your answers can help you better reclaim your time from gaming by giving you specific insights into your motivation and showing you reasons for your choices you may not have considered. The pursuit of self-improvement can have benefits besides dealing with unwanted behaviors. If one of those outcomes is that you know yourself better, that's a big win.

Chapter 16

CULTIVATING CREATIVITY

FOR MY NEXT TRICK...

One of my hobbies is magic. I started performing up-close and stage magic as a teenager and never stopped. Although I often like to open my seminars and keynotes with an illusion to illustrate the topic at hand, I enjoy studying techniques and "secrets" of magic even more. There's no other discipline like it.

When I studied math and physics, I learned a great catalog of principles and techniques, then applied those approaches over and over to various situations. But when it comes to magic, there are few principles that get used repeatedly. Nearly every trick requires a unique method; for instance, the technique behind making a coin disappear is good for only that and not much more. The same goes for making an object float in the air, mindreading, and sawing your assistant in half. The science of science calls for a consistent approach used repeatedly. The science of magic calls for a stream of one-of-a-kind ideas used for a single purpose and probably nothing else. To create a new magic trick is to engage in profoundly creative thinking, because you have to blaze your own trail, and likely think in a way that no one has before.

People who design original magic use the brain in ways that—ahem—most people aren't creative enough to imagine. Magic demands constant invention because every trick is a new puzzle, a fresh challenge to the ingenuity of the mind. In this way, the work of designing magic is also the dopamine system firing on all cylinders: it requires serious creativity.

HOW TO BE CREATIVE WITHOUT TRYING

Creativity is in many ways an unconscious act, but there are things we can do purposefully and consciously to increase our creativity.

You've heard the old saying that "life is what happens while you're making other plans." In the same way, creativity is what happens when you're doing other things—and dopamine is all about bringing us into contact with other things. That's why people who rely on their creativity for their art or their employment often carry a notebook. When an unusual thought occurs to them, they write it down. It could be a different perspective on an old idea, a combination that hadn't occurred to them before, an unusual phrase, a provocative image, a striking turn of phrase, a word that sounds strange—and none of these arrived by looking for them. This is how creativity often works: the unconscious mind sorts through the otherwise low-salience stimuli hanging around inside us, weighing raw input and promising combinations for novelty and utility. The output might be fully formed or not. It might be something to prompt further thought later. Whatever comes out, our minds are doing divergent and convergent processing in the background, then offering up a potentially useful result as it appears.

We get far fewer original ideas from "thinking harder" than we do from letting our minds wander. Illustrator and graphic designer Christoph Niemann says, "I hate so much the idea of no control, but this approach of not planning opens a new door. It's really hard, but it just leads to these magic moments."[125] We can invoke Dorothy Parker's "disciplined eye" at will. It's the wild mind that comes more often unbidden. Creative thinkers covet those moments.

Breakthroughs not just in art but also in engineering, mathematics, language, and science come when creative minds take paths that no one else has followed. Wandering from the well-worn path happens most often when we let our minds wander to other matters—ironically, when we're not trying to be creative at all. By being aware that creativity arrives on an uncertain schedule along unexpected roads, we can be on the lookout for useful material and capture it—in a notebook, a text, an email, or something else. Being creative is just as important as noticing when you're being creative and not even trying.

PRACTICE!

You've heard the old joke: How do you get to Carnegie Hall? *Practice, practice, practice!*[126] Turns out that, in one key way, creativity is a skill like playing baseball or speaking a foreign language: the more you do it, the more accomplished you get.

Creativity is built on connection. Isaac Newton connected the acceleration of an apple falling from a tree with the orbits of the planets around the sun. Charles Darwin developed the theory of evolution by connecting his observations of species variation with James Hutton's ideas about changes occurring over geologic time scales. Making connections in the world seems, well, *connected* to connections within the brain. Researchers estimate the strength of these neural connections by measuring whether activation of one brain network leads to simultaneous activation of the other. Think of it like figuring out which cord on the power strip is the one connected to your computer: you jiggle it at the laptop end and see which one moves at the other end.

One way to test the strength of these neural connections is the Remote Associates Test. This protocol measures how well a person can make connections between seemingly unrelated words. Each question presents three cue words that are linked by a fourth word, which is the correct answer. Sample items include *safety/cushion/point* and *pie/luck/belly*.*

A group of neuroscientists found that people who get high scores on tests of creativity and who are successful in creative fields have strong connections among three networks within the brain.[127] We're already familiar with two of the three regions, the desire and control dopamine pathways. The third region is called the *default mode network*, which becomes active when we're daydreaming. These three networks likely promote creativity this way. First, the default mode network produces daydreams. Its strong connection to the desire circuit delivers whatever novelty dopamine is detecting, and that too may become a part of a daydream. The control pathway then evaluates the resulting ideas and selects the ones that are potentially useful. The researchers were right: strong connections make strong creativity. Repeated activation of brain circuits makes the connections stronger among them. "Cells that fire together wire together" is how psychologist Donald Hebb described it. That's why when you practice something, it becomes easier to do, including creativity. When another group of scientists tested this idea with creative people, they saw that it was true. As professional pianists improvised jazz music, the researchers saw specific regions within the cerebral cortex lighting up. When they measured the level of connectivity among these regions, they found it was directly related to how much time a particular pianist had devoted to improvisational practice.

Creative ideas often feel like they come out of nowhere, but there really is a source. They come from regions within the brain that we can't directly access. Still, we can make use of it. Although we can't often summon a creative solution to a difficult problem on demand, we can make its appearance more likely. All we have to do is practice. If we strengthen the neural connections associated with creativity, there's a good chance our efforts will pay off.

If you want to build your creative capacity, start doing some creative thing frequently: draw pictures, write stories, figure out story problems

---

* The correct answers are "pin" and "pot."

in a math book, work puzzles. Building your brain connections will build your creative ability. To return to that old joke, if you want to get ahead, you've got to practice, practice, practice.

REDECORATE AND REFRESH

Another way to build your creative capacity is to surround yourself with inspiration.

Customize your office, studio, or workspace. Cover your walls with things that comfort you and challenge you at the same time. I've done this in my office. The walls of this former one-car garage are crammed with framed pieces from floor to ceiling. In some places the frames are so close you can barely see the wall between. I have about two hundred objects ranging from autographs and antique political buttons to full-sized movie posters. One of my most prized possessions: an original poster from the movie *Midnight Cowboy* complete with the X rating in the lower left-hand corner. I also have autographs from Bob Dylan, Don Knotts, Robert De Niro, and Vince Gilligan, and a photo in front of Broadway's Belasco Theatre of me with Bryan Cranston, the actor who played Walter White on *Breaking Bad*, a show whose moral questions will always draw me in. Wherever my eyes fall, I get a happy memory and a little creative provocation, plus a blast of inspiration from my creative heroes. They followed their dreams. So can I. Todd Lubart, professor of psychology at the Université de Paris, said that while we all take inspiration from the world around us, "some people use better grade fuel."[128]

Explore what you think doesn't matter. Creative people aren't picky. They take in what others pass by, even when it doesn't appear to be relevant to their work. I advise my students, "If you want to become a better writer, learn a bit of a new language. Read an article about something new. Spend an afternoon at the museum looking at art." Refill your stock of potential creative elements as an ongoing act of divergent thinking. Seek out what you've ignored. Open your mind.

Engage with art. A study done in South Korea found that when volunteers were asked to generate creative ideas for business, such as designing a computer keyboard, naming a pasta brand, or thinking of new ways of recycling, those who had viewed art prior to the exercise generated a larger number of ideas and their ideas were more original (i.e., less obvious).[129] There is power in pretty things.

Get into nature. Although cities present us with creative architecture, street art, dynamic advertisements, and dissonant experiences for our senses, natural environments seem to enhance creativity even more than those things.[130] Perhaps nature restores our capacity to pay attention.[131] That's important because purposeful divergent thinking is fatiguing. After a while, we run out of energy for it. Spending time in a natural environment is less demanding, yet the diversity of what's available there is still a source of new material.

Once something captures our interest, paying attention to it no longer requires effort. This is called *fascination*, and it's why children with ADHD can spend hours focusing on a video game but become distracted after only a few minutes of homework. All kinds of natural phenomena are inherently interesting to human beings, but nature doesn't monopolize our attention—call it a "soft" fascination because it leaves room for reflection and restoration. The "hard" fascination of something that absorbs us completely is more experiential than reflective. Nature doesn't require our concentration. The familiar, even gentle stimulation fairly invites reflection, which is the path to new perspectives—to creativity.

Not everyone has a park nearby, and that's okay. Try background recordings of wind, rain, or natural sounds. Even pictures of nature[132] seem to help, as do plants.[133]

USE YOUR DREAMS

Aldous Huxley advised us to "dream in a pragmatic way," meaning we should tailor even our biggest plans to fit the possible. But the philosophy that dopamine leads us to is less like Huxley's and more like

CULTIVATING CREATIVITY

George Bernard Shaw's, who said we ought to "dream things that never were"—a dopamine specialty—"and say, 'why not.'"*

When you sleep, your brain operates differently than it does when you're awake. The levels of H&N chemicals go down and dopamine goes up. This can be useful for anyone trying to be more creative. While we are not closed off from our senses during sleep, they become less important to our mental processing than dopamine, which can now make uninterrupted connections among whatever is already on your mind or hiding away in memory.

For most of us, this unimpeded dopamine experience is the closest we will come to experiencing aspects of mental illness. We know this is true because when we compare the dream stories of healthy people and the waking stories of those with severe mental illness, there is no difference. It's not a surprise: many of the pathways of healthy creativity also activate during schizophrenic experiences. Art, especially music, is full of examples of the result of overstimulated dopamine circuits, whether brought on by mental illness, drugs, or both. While drugs are risky and mental illness is painful, some of the creative products they've produced are beautiful, amazing, and moving. But you don't have to get high or suffer mental illness to stimulate yourself toward creativity. Dreams can inspire anyone.†

Episodes of dreaming occur throughout the night, with the first episode usually beginning around ninety minutes after you first fall asleep. Dreams are often characterized by bizarre, irrational material, but dreaming isn't the only time during sleep that our minds generate

---

* Shaw didn't say this off the top of his head. It's a line of dialogue from his 1921 stage play *Back to Methuselah*, a three-hundred-page journey from Biblical Creation to the year AD 31,920. The line is spoken by Satan to Eve in the Garden of Eden, and it comes on the first night of a play that takes three nights to perform. With all that in mind, perhaps we should take it as a warning, not encouragement.

† Some kinds of mental illness are at times associated with great creativity, but they are not conditions to leave untreated, nor is this something to aspire to. The profound suffering that can come with mental illness makes creativity more difficult to exercise, not easier. Treatment is available. If you need help, get it. You will feel better, and you deserve to. Your art will be waiting for you on the other side and it will be better under your command.

bizarre ideas. It also happens just as we're crossing the threshold from wakefulness to sleep. This twilight of consciousness is when we sometimes get our best ideas. It's called *hypnagogia*.

As we're drifting off, our thoughts tend to become muddled. Reality twists and warps, mingling with our own imaginings. The mind travels from one idea to another more fluidly, often along irrational pathways. If you've been focused on a specific activity before going to bed, that may dominate the experience as you drift off. This has been called the *Tetris effect* because people who play the video game Tetris for extended periods often find themselves compulsively thinking about how shapes in the real world can be rotated to fit together.

If you've been working hard on a problem before you go to sleep, the hypnagogic state may produce divergent, solution-oriented thoughts on the matter. Usually, however, the state slips by unnoticed. The exception? When we happen to wake up in the middle of it. American inventor Thomas Edison (who acquired over a thousand patents) found this state to be so productive, he came up with a clever strategy for ensuring he would retain the ideas that were generated. He would settle himself into a comfy chair and close his eyes to go to sleep. In his hand, he would hold a heavy metal ball. Muscles relax in the hypnagogic state, so when he reached it, the ball would fall to the floor with a loud bang and wake him up. That allowed him to jot down any creative ideas that had emerged during the hypnagogic state. The surrealist painter Salvador Dalí did something similar using a heavy key. One can easily imagine the dopaminergic state of hypnagogia being of great value to a surrealist.

Scientists tested this strategy in the laboratory by giving volunteers math tests.[134] All the problems could be solved through a straightforward though laborious method—but there was a sneaky, easy solution too. After working on the problems for a while, half the volunteers were instructed to take a nap. The scientists monitored their brain activity so they could wake them during the hypnagogic state. The volunteers reported strange visions that seemed to be loosely related to the research. They described "dancing numbers and geometrical shapes, the Roman Colosseum, and a hospital room with a horse."

Among the volunteers who took the interrupted nap, 83% found the hidden key to the math test compared to only 30% of those who stayed awake. The effect disappeared if a volunteer was allowed to enter a deeper stage of sleep. One of the researchers commented, "Our findings suggest there is a creative sweet spot during sleep onset. It is a small window which can disappear if you wake up too early or sleep too deep."

Among the nappers who found the key, the insight didn't happen right away. The solution tended to dawn on them after they went back and spent a considerable amount of time wrestling with the problems. It's reminiscent of the luck of the prepared. The desire dopamine system doesn't deliver on command, but this research suggests there are ways we can work with it to make it an ally.

When you wake from a dream, you're conscious at the same time dopamine is the most powerful force in your brain. Your thinking is more fluid, you're making leaps from topic to topic, and logic doesn't matter. This can be the most creative moment of your day. The trick is catching the memory of those dreams before they fade away. To take advantage of this kind of creativity, make a plan to make use of the moments just after you wake up, a time we referred to in *The Molecule of More* as "this crack between two worlds."

Focus deeply on a problem of interest just before you go to bed. The brain continues to work on problems while we're asleep. By making a problem the last thing you think of as you drift off, you're "assigning" the task to your unconscious mind, a powerful source of original thinking and new ideas. Have you ever had an answer just come to you out of the blue? That was a product of unconscious processing. "Sleeping on it" is a way to access this power.

Put a notebook by your bed. When you wake up, write down what's on your mind.

Write it down right away. Don't simply say you'll remember it later. You won't. In those moments that your brain isn't fully engaged in conscious thought, your memory isn't either.

Don't judge it for value. Treat it as what it is, a contribution to divergent thinking.

Don't try to make sense of it right away. It's probably symbolic and metaphorical. Analyze it later.

Decide to be open to the unlikely. Let dopamine do what it does best.

DON'T WAIT FOR SOMEONE TO GIVE YOU THE GO-AHEAD

As you learn in elementary school, two plus two is four. And it's going to be four every day, all day. You can swear it's five and that won't change a thing. But sadly, facts don't matter in every situation. Comedian John Mulaney says, not entirely joking, that a lot of getting through college "is just your opinion." In a literature class, he says, you might raise your hand and say, "'I think Emily Dickinson's a lesbian,' and they're like, 'Partial credit!'"[135]

When it comes to predicting who will be creative, researchers rely a lot on "partial credit." That's because in a few areas of "knowledge" there aren't that many facts to have—at least facts that are useful. Since we know more about the ingredients of creativity than its process, tests of creative potential are usually just subjective, even opinion-driven inventories of qualities. That's a limited approach. Saying, "This person lacks certain characteristics so they're less likely to be creative" is like saying, "This car has no gasoline engine so it's probably not going to work." *Tesla, anyone?* Thinking about creative potential as ticking the required boxes is a poor way to approach the matter.

Consider one of the most widely accepted tests of creativity, the Torrance Test of Creative Thinking (TTCT). It focuses only on divergent thinking. The reasoning phase, convergent thinking, goes begging, yet the TTCT is among the most widely respected and widely used of these exams. It tests for five abilities:[136]

- Fluency, the creation of multiple ideas relevant to the first
- Originality, the creation of original ideas that are statistically rare

- Elaboration, the development of additional related ideas
- Abstractness of Titles, the creation of abstract ideas to convey the essence of a category
- Resistance to Premature Closure, the processing of new ideas on a sustained basis

If you took the test, would scoring low on one or more of these discourage you from trying to be creative? After all, it's a test created by experts. Shouldn't they know? But wait—before you decide you're not creative, consider this: earlier versions of the test measured only four of these abilities. You might have come out as "creative" on that version and feel great about yourself. Too bad you took the newer version. *Hmm.*

Perhaps you could see how you do on other tests. There are quite a lot, such as the thirty-four yes–no items of the Biographical Inventory of Creative Behaviors; the Creative Product Inventory and its categories of Generation, Reformulation, Originality, Relevancy, Hedonics,* Complexity, and Condensation; and the Creative Product Semantic Scale with its three axes of Novelty, Resolution, plus the grab-bag twofer of Elaboration and Synthesis. Oh—and be sure to bring an extra pencil, because educators and psychologists typically assess creativity with an intelligence test, too, such as the Stanford–Binet.

Make no mistake, these tests are valid investigative tools. They measure what they say they measure and do so accurately. They have their uses. The problem is attaching the word *creativity*. That skill is simply too big, too multidimensional, and too beyond our imagination to be dissected "by the numbers" *at this point in time*. The brain in many ways remains a mystery, and creativity is an especially challenging process to quantify.

---

* Let me save you the trouble of looking it up: hedonics is the ethical consideration of the relationship between duty and pleasure. It's also the branch of psychology dealing with the study of pleasant or unpleasant sensations.

The beautiful thing about creativity is that it is infinitely pliable in appearance and definition. That's good news for those who appreciate creativity, but for anyone who wants to measure it, it's a stopper.

Here's the point: Start being creative right now. Don't wait for some test (or, God forbid, some "expert") to predict whether you'll succeed. The ability to be creative is in every one of us. It requires only exercise and cultivation.

As a university professor, I meet lots of people who are certain that once they are equipped with a formal imprimatur, they will be able to start in on what they really want to do. That's true for electrical engineering but not for aspiring painters, musicians, writers, comedians, philosophers, landscape artists, and so forth. It makes me sad to meet people who need external validation before they will pursue the creative life they want. For every great singer coming out of years of classical training, there are dozens more who got good at it by doing it for fun on their own. For every novelist who spent a couple semesters at the Iowa Writers' Workshop, there are hundreds who decided they'd like to write a novel instead of just read one, opened a Word document, and started filling the pages. Training is fine but it's not necessary. Creativity itself is already inside you.

If you want to be creative, you already believe that it's going to be fun. That's reason enough to get started. If you find the work isn't for you, it's a safe bet that a certificate won't change how you feel. Personally, I play guitar because I like it, not because I'm good at it. Creativity is a delight no matter how good you are at the start. Aptitude doesn't matter, only desire.

BUILDING A CREATIVE STATE OF MIND

You know the quote: Any sufficiently advanced technology is indistinguishable from magic.[137] It's the same for creative power. Creativity is the product of dopamine circuits in the brain, largely hidden and mostly inscrutable. Inspiration arrives unbidden and without warning, as if the universe has blessed you with the answer. Even when we know what's really going on in terms of neuroscience, the feeling remains of having experienced a miracle. The good news is that we understand enough about creativity that we can make miracles more likely.

It begins with the cultivation of a state of mind favorable to the process. We must use our conscious mind to make our unconscious mind aware of what we'd like to figure into the answer we seek: our likes and dislikes, limits, sensibilities, skills, long-term goals, short-term desires, and what we're willing to accept in the end. From that, the creative process proceeds as dopamine-driven speculation about the future using all those building blocks of the here and now. In short, our ability to creatively imagine the future depends on our ability to interact with the present. We assemble what might be out of what is and derive order out of disorder. This is how humankind acquires what we did not have before: new, highly valuable, often life-changing insights, methods, and technologies.

You will be more creative if you emphasize the gathering phase as a separate act from the analysis phase, and if you improve and expand your catalog of other abilities. The methods and encouragements in this chapter are all things you can easily make a part of your life. Also, unlike some of the techniques for addressing other kinds of dopamine-related challenges, the things that build creativity are almost all easy and fun. The most effective thing you can do to be more creative? Practice new ways of thinking in everything you do:

Give it all you have. Louis Pasteur, who discovered the principles of vaccination, famously said, "Chance always favors the prepared mind." Original ideas don't appear out of nowhere. They come from a vigorous approach to life and a mind filled with facts and understanding. It's been called "the luck of the diligent."[138] So when you need to learn something new, don't learn it halfway. Learn it completely. And when you work at something, don't give it half an effort. Give it all you have.

Seek out whatever is different, unusual, or out of place. Our sense of salience directs us away from anything that seems less than useful, but we can override this habit with the conscious mind. One aspect of embracing the here and now is giving purposeful consideration to the things around you, especially those things that you've typically ignored. Start looking for them. Break the tendency to overlook ideas that fall outside the expected domain. One reason that fortune favors the bold is it takes a bold personality, or at least a highly confident one, to entertain ideas from any source.

Look for novel uses for ordinary objects. This can strengthen the connections between diverse regions within the brain and make you more creative overall.

Occasionally work on several things at once. As we've seen, switching between several tasks rapidly shakes you free from cognitive fixation. Scheduled task-changing is more effective than changing on your own. The otherwise inefficient approach of multitasking can shake you out of your usual ways of thinking and open you to new habits and, hopefully, fresh, creative answers.

Switch sides. Artists gain creativity by growing their capacity with quantitative reasoning, and logic-minded people gain creativity when they apply reason to what seems subjective. There is often exquisite order within chaos, and intricate beauty within order.

Plan times to let your mind wander. When your mind is unfocused, it's kind of like sleeping while you're awake. In moments like that, dopamine has more power and the overwhelming here and now fades. This is when creative ideas are more likely to appear. Do something aimless, or tuck into some mindless work like clearing your inbox or sorting a drawer. The more you do this, the more your creative side can root around in the latest delivery you've provided from the here and now.

Creativity arises out of a rich interior life. The only one who can build your mind this way is you.

Finale

THE NEED FOR MORE THAN MORE

In Conclusion
FINDING A WAY TO LIVE

It defines our lives, this emotional buildup when we want something and the letdown after we get it. The seemingly inseparable feelings of anticipation and emptiness define the perpetual, and perpetually frustrating, experience of the dopamine chase.

Victory is fleeting. Expectation always exceeds outcome. That's why, for each of our pursuits—of romance, career, material things, short-term pleasures, long-term achievements—we look for ways to mitigate the anticipation and lessen the letdown. It's been the point of this book. We need ways to proceed more happily and constructively through life.

But there's a bigger problem. A trip through the dopamine cycle plays out over a few seconds, minutes, days, or weeks, but it reflects the larger experience of a life itself. Over the years of pursuit and achievement, we begin to wonder what it adds up to. *Why are we here? Is that all there is?*

There ought to be a better answer than to cobble together our comforts, assemble a set of distractions, or pretend that explaining why the brain works this way is the same as making us feel better about an existential void.

Is there some animating truth behind what we're doing with our days? What's that highly anticipated future for, anyway? What are *we* for?

How about an answer for that?

THE SOUND OF A TUNING FORK, STRUCK UPON A STAR

WONDERFUL, DEMANDING HOPE

> *His heart beat faster and faster as Daisy's white face came up to his own. He knew that when he kissed this girl, and forever wed his unutterable visions to her perishable breath, his mind would never romp again like the mind of God. So he waited, listening for a moment longer to the tuning fork that had been struck upon a star. Then he kissed her.*
> —F. Scott Fitzgerald, *The Great Gatsby*

Fitzgerald knew what happens when *unutterable visions* yield to *perishable breath*. Hell, he wrote a whole book about it. You drag your dreams to the window and bet them on the promise of some perfect reality, understanding that only when the transaction is made will you fully know the futility of the bargain. Gatsby went ahead anyway, though he paused to savor the anticipation one last time,

listening for a moment longer to the tuning fork that had been struck upon a star.
What is real and in your arms is no match for what you imagine. You fell for it again.

Hope, as they say, floats. We carry it weightless within, and in that way it carries us too. Hope takes us toward something we want, can take us right up to the thing, but it has to leave us short of the goal—because hope doesn't deal in having. Hope is fuel to get you there, nothing more, and it pushes us forward by encouraging us to believe perfection might be out there instead of reality, which is corporeal and therefore finite and lesser. Hope is the glory of maybe. Possession—reality—is the end of that. When you possess what you desired, you are confronted with the difference between what is and what could be. If your mind is going to romp again like the mind of God, you're going to have to start again, maybe on the last wish one more time, maybe on another.

And that's what you'll do.

We are bound to look forward; our minds are the captive of our biology. We are suckers for the twinkling perfection in a star too far away to touch. And we'll run toward it, and that is what will take us to tomorrow. Another Daisy will inspire glorious, unutterable visions, and then she will arrive, and we will savor her white face as it rises to our own, and we will think about the cost of that kiss—but we won't, not really, because we hold out that ridiculous hope—*hope*—that this time the reality will match the imagining. We'll muse over the possibility, and that will be enough to convince us one more time to trade our unutterable visions for her perishable breath. But the transaction won't do any of what we imagine. Tricked again. It doesn't take many of these letdowns to make us cynical toward the idea that there is meaning to any of this.

But I say there is a way to be satisfied. It's not difficult but it demands self-discipline, to bring meaning to our day-to-day efforts, including when we struggle between the invitation of hope and the bruises of reality. To do so, we must first identify virtues that mean something to us, that we consider bigger than our own desires. Then, and this is the demanding part, we must choose actions and responses across our lives that are in line with those higher purposes. If we do this, we will see meaning in our struggles, and thus in our lives. Again: We can use our choices to promote not just *more* but some *better* more, a point on the horizon that matters to us and to humankind at large. In this way we will discover joy in what we now dismiss as a damnable cycle, a yin and yang each time satisfying only for as long as it takes to find out it is not. "I was within and without, simultaneously enchanted and repelled by the inexhaustible variety of life," *Gatsby*'s narrator Nick Carraway says. But when we identify a higher purpose, one we choose, we impose meaning on what otherwise has no meaning. This is the way through.

BOATS AGAINST THE CURRENT

There is no more moving, poetic, and depressing portrayal of ambition as the off-ramp to emptiness than F. Scott Fitzgerald's *The Great Gatsby*. Born into the middle class, Fitzgerald ascended to high social standing in part by his courtship of Chicago heiress Ginevra King (the likely model for the character Daisy Buchanan), then through his marriage to Zelda Sayre, a wealthy Southern debutante whom he met while he was in the Army. For more than a decade Fitzgerald lived the high life, enjoying his literary reputation and gadding back and forth to Europe, while struggling like mad to pay the bills with his writing. He lived feast or famine, always on the hunt for the next big check. He finished *Gatsby* in October 1924 while he was stretching every dollar he had to maintain himself on the French Riviera. (He had a spectacular deal: the villa he rented was $79 per month, about $1,300 today.) By the time he left, he had almost no money in the bank, though that was also what he had when he arrived. One of the next things he wrote was an article for *The Saturday Evening Post* called "How to Live on Practically Nothing a Year."

The constant uncertainty of his finances must have depressed and motivated him at the same time. He knew how the business of writing worked. When the tastes of the public coincided with his art, the financial rewards and cultural recognition flowed. Yet he never seemed to pause, didn't stop working for a while to enjoy the fruits of his effort, or reflect and adjust his course after a win. In his famous novel, Fitzgerald presented the green light at the end of Daisy's dock as a promise and a lie. It calls us to the thing we desire with all our hearts, a thing we may chase without even knowing what it is or what we will do with it when we get it. It calls us to our ultimate happiness, but according to Fitzgerald there's no such thing. As Nick Carraway puts it in the end,

> *Gatsby believed in the green light, the orgastic future that year by year recedes before us. It eluded us then, but that's no matter—tomorrow we will run faster, stretch out our arms farther . . . And one fine morning—*
> *So we beat on, boats against the current, borne back ceaselessly into the past.*

As lovely as that famous passage is, it's only nihilism with a bow on—a bruise for show, a "pretty war," as songwriter Jeff Tweedy wrote. And as a way to live, it is nothing more than lazy. You can either do the heavy lifting it takes to find meaning or you can pose with the pain. It's as if Fitzgerald mistook hopelessness for nobility. Did he never look around as he rose in his world? Did he never consider that a lifetime of wins and losses could add up to something?

Our serial victories and struggles leave us unfulfilled. Well, *duh*. It is a complete misapprehension of life to cast human experience as systemic failure, though it's good for a poetic account of the obvious. When we get what we want, we do not stop living one way and start living another. When we secure the prize we sought, we get bored with it pretty fast. A child on Christmas morning wants to tell you about his new toy. A child on Christmas afternoon will ask you what's next. What we need is a unifying purpose behind it all.

Here's the way: The pursuit of meaning via our day-by-day achievements gives us happiness along the way, and if our achievements are toward some larger purpose, then each no longer leads us back into emptiness. We deserve to take pleasure in doing well the things that we do: in learning, in striving valiantly, in the self-improvement that arises from error and shortcoming—because we do them for reasons beyond our own comfort. In this way the component pursuits of the larger effort acquire value too. And thank God for that, because they take most of our time. We need a way not just to pass the time but also to enjoy it, and toward some better and greater and more lasting end. This is it.

Happiness is pleasure in the activity of the moment. Fulfillment is a sense of purpose as life proceeds. Happiness is joy in the here and now. Fulfillment is joy that there is more than the here and now—not mere optimism but confidence, even certainty, that we are creating something of lasting value. We tend to believe that it's an either-or choice: you can be happy now and let down later, or suffer now and be fulfilled later. At the biological level, where we began this discussion so many pages ago, it seems that happiness and fulfillment are opponents—because they are. But it doesn't have to be that way. There is a way to live that is richer than what we may think is possible. It comes when we organize our choices to make partners of happiness and fulfillment. We're better off with both.

Here's something else, too—something bold: it's impossible to consistently experience happiness without fulfillment. Each experience is more intense when the other is also present.

A LESSON FROM DAD

A life of purpose equips us to better deal with pain.

For his whole adult life, my father was a country preacher whose appeal behind the pulpit came from a combination of born-to-it charisma and a poet's capacity with words. He was beloved by his congregations. But about halfway through his career he experienced a crisis of faith and left the ministry. For five years he drove for a private ambulance service. Before cell phones or pagers, every call came in by landline to a number that rang at the office and at our house, and it had to be monitored 24/7. When he was home he had to stay close enough to hear the phone. This greatly restricted our ability to go much of anywhere beyond the backyard, but I don't think dad minded all that much. He was withdrawing from the world.

The service operated as an adjunct to a funeral home, a common practice at the time. The ambulance itself was one of the funeral home's two hearses. To make one into an ambulance, Dad would attach a flashing red light to the roof. When I was about eight, I remember asking my father a ghoulish question about the arrangement: Since the funeral home makes money providing funerals, is that really the operation people want in charge of getting them to the hospital in time? He explained to me that while some people might prioritize profit over life, a decent person will not. It is up to us to do the right thing. When nobody else is watching, the obligation is ours alone to fulfill.

This was one of the first "grown-up" facts of life I understood. I couldn't have put it in these words at the time, but I was learning that a life worth living must be more than a series of transactions toward happiness in the moment.

THE ARC OF THE UNIVERSE IS UP TO YOU

Neither morality nor meaning is built into the nature of things.

Contrary to the famous quote, the arc of the universe does not bend toward justice. It bends toward personal ease and chaos. Goodness, order, and kindness arise for one reason alone: because we choose to do good things for others, sometimes at great personal sacrifice. We do these things in opposition to the lazy entropy of human nature. In so doing we attach ourselves tighter to community and humanity—or, as my father would have said from the pulpit, to instill some virtue in ourselves. This, he told me, was what we are wisest to seek in life, a trait of God Himself.

As for how that pertains to the problem we're addressing in this book, try this. When we do what we do best and enjoy most in service to the virtues we love, pleasure both in the moment and in the long run will usually follow. This is the kind of happiness, both Aristotelian and modern, that will sustain us. It is the only happiness sustainable throughout life. The pursuit of meaning gives us the strength to persevere, to give of ourselves, and finally to face the end of life.

The cause of my father's crisis of faith was the loss of a child during birth. Dad suffered a blow to the solar plexus of his sense of self; it shattered a beam that supported his life's meaning. I cannot be sure, but here's what I think happened next. It took dad five years of driving that lonely ambulance to see that, when it comes to life, pain is so much a part of the arrangement that we cannot live in it long enough to inure ourselves to the ache. He knew because he tried. He spent half a decade focused on the most heartbreaking part of that rural world where we lived, racing to car crashes and heart attacks and stabbings and desolate human endings. He even tried denying himself his family, tethering himself to a telephone. Finally he realized that none of this diminished the loss of his baby. He died before I had considered asking him about all this but I believe I know now what he would have said: *You can't make pain hurt much less but you can find a place for it.* Profound pain gives us a binary choice, to use it for good or to be consumed by it. There is no in-between, and there is only one constructive way out: to let it remind us to direct our attention without instead of within. My father spent five years on the wrong approach before returning to the right one, when he came back to the pulpit to look less at himself and more at others, as before.

The great Dutch theologian Henri Nouwen put it this way: "We human beings can suffer immense deprivations with great steadfastness, but when we sense that we no longer have anything to offer to anyone, we quickly lose our grip on life."[139]

When it comes to problems of love, sex, ambition, and the rest, the idea is not to transcend their challenges but to experience them in order to understand their normalcy, and then to put that understanding to good use. We survive and grow and even thrive through each trial, each obstacle, not by focusing on the branch but by returning, relentlessly, to the root, which is meaning. If we are to be satisfied, even happy, in the long run, we are obligated to consider the full meaning of life, to find a way to articulate it so that we can say it out loud, and to apply it in every situation.

This brings us back to the start, the dopamine chase, and our goals: the easy goal of pleasure in the moment and the elusive goal of satisfaction in the long run. When meaning is a part of the cycle, the emotional experience is transformed. Each victory in a dopamine-inspired pursuit gives us three things. It helps us infuse with purpose the choice of challenge we ought to take up next. It advances our larger, lifetime purpose. And because we pursued the challenge in the spirit of a larger meaning, victory gives us a reminder to reflect on the richness—the joy, the appreciation, the satisfaction—we felt in the effort. In this way we enjoy life more consistently and moments themselves more deeply. This brings me to the most comforting lesson I know.

My cousin, college roommate, and close friend, Kent Northcutt, died from complications of diabetes at age thirty-eight. He left behind a wife, three small children, an interesting career, and bereft friends who turned out in such numbers that the crowd spilled out of the huge church into the street. A pastor, Chris White, shared something that day I've returned to over and over. He said we might not remember much of the time we have with each other, but that's okay, because "it happened anyway." We weren't doing things together so we could remember them later. We were enjoying the experience of being alive. Seize that. It is an especially rich kind of plenty.

AN ANCIENT APPROACH TO FINDING MEANING IN YOUR LIFE

Aristotle said that pleasure is not a state of mind but a byproduct of doing something—carrying out, not just being. We have this experience most intensely when we do something we're suited for. That makes sense. I won't get much pleasure from drawing because that's not my thing. But I'll get plenty of pleasure from an afternoon playing backgammon because I like the strategy and the challenge, even when I lose. I'm suited for it, but what does it mean, to be suited for something? It means we're skilled at it or that we enjoy the effort it takes to gain that skill. (Aristotle connected learning with pleasure as well.)

But there's still an element missing: the activity I choose, if I invest deeply in it, ought also to advance a higher goal. Could backgammon be aligned with my higher purpose? If it makes me better at strategy and critical thinking, that could be a reasonable alignment. But let's consider more obvious connections. If I'm *suited for* engineering, the time I spend doing it produces not only a period of happiness for me but also some outcome of use to someone else. If I'm suited for plumbing, I take pleasure in my effort while making life easier for people who have problems with their pipes.

Those results of my "pursuit of happiness"* cascade into something more robust and long lasting than fun—which is a pretty good outcome, considering the original goal was only to pass the time in a pleasant or profitable way. By doing that for which we are suited, we create value for others. This in turn contributes to the satisfying feeling that our life has meaning—that it has positive impact beyond ourselves. And this is how Aristotle's powerful idea applies to the problem of the dopamine chase: pleasure and meaning are experienced consistently only when they are bound up together.

So what do we do in particular to make this into reality?

To create a life with meaning, pleasure, and peace with ourselves—a life of *happiness*—we must, Aristotle said, find the *appropriate* pleasures for our life. What did he mean by "appropriate"? Aristotle described this as any pleasurable activity that elevates *virtues*: qualities of character or moral habit that become our guide, our touchstone, for action and reaction. In this way we can grow to more reflexively do the better thing. We are also freed from having to address each challenge as it arises and think through the same issues over and over again.

Aristotle proposed four primary virtues: **wisdom**, choosing to do

---

* From the Declaration of Independence: "We hold these truths to be self-evident, that all men are created equal, that they are endowed by their Creator with certain unalienable Rights, that among these are Life, Liberty and the *pursuit of Happiness*" (emphasis added). Thomas Jefferson borrowed the phrase from political philosopher John Locke, who was greatly influenced by Aristotle and who used the word "happiness" in the Aristotelian sense of pursuit of purpose. Today we tend to see "pursuit of happiness" as an out-of-the-blue message from the past to remember to go have fun. In this book, I too have used the word "happiness" to refer to pleasure in the moment. But Locke and Jefferson were referring to Aristotle's assertion that a life of meaning, of *happiness*, is one in which we pursue activities and their attendant pleasures to advance virtue.

the right thing; **justice**, dealing with other people fairly; **temperance**, avoiding the extremes of self-indulgence in the pursuit of pleasure; and **courage**, doing the right thing even when we are fearful.

Aristotle decided that the primary virtue he would pursue was wisdom. This brought him both pleasure in the act of pursuit and meaning in the result, since he shared his wisdom with others, fulfilling his higher purpose. As a philosopher, Aristotle's choice of wisdom as his primary virtue was easy. The rest of us have some thinking to do.

What do you consider virtuous? What larger, selfless pursuits move you most? Few of us have thought deeply about that. When we do it's usually as a reaction, not a focused consideration. We most often choose from ready-made options presented by family, friends, culture, and religion—we borrow our opinions instead of working them out from self-examination. But to live a life with meaning that matters to us, we have to choose purposefully so we can embrace our choice fully and sincerely. We have to have a solid answer to *why* in the virtue we choose. This is the moment when religion, philosophy, psychology, politics, and your own experience come crashing together, and only you can sort through them. What are the pillars of your outlook? What virtues do you look to as first principles to make the choices in your life?

When you choose the virtues in life that will be your aims, you equip yourself to find meaning in more of your actions. Aristotle said we should begin with what we do well, to ensure that we act with enthusiasm and experience motivating pleasure. From those things, we then choose to pursue activities that advance our purpose. We'll always have obligations that divert us from this ideal path, but if we start with those activities that give us pleasure while leveraging our skills and supporting our elect virtues, we'll be building our fulfillment in the moment and across life.

Dopamine pulls us to the immediately pleasurable. The tools in this book can give you the power to fight back. But these techniques alone will not lead to consistent satisfaction over the course of an entire life. What matters is the meaning behind it all. You'll be better able to experience pleasure for its own sake, plus this pleasure will signal that you are pursuing an activity with value beyond yourself. Remember that dopamine promises pleasure, not meaning. It is only we, with our capacity to love, that give meaning to the indifferent universe—but we must find it by explicitly attaching to our days something beyond ourselves: wisdom, justice, temperance, and courage, or a god that calls you to these things.

When we do, we will better remember that pleasure is a brief experience, not a complete purpose for life. We gain a way of navigating life *toward* meaning *by* pursuing activities that bring us appropriate pleasure. Instead of life as a series of desires, successes, and letdowns, life becomes a journey, with pleasure in a place where it has value because we use it to identify the best activities for us and to amplify their positive effect for ourselves and others. By explicitly thinking about our aspirations and documenting them, the choices we make in our daily lives will be better informed by the virtues to which we aspire—not perfect, not always complete, but better. Our lives will embody a purpose beyond ourselves. As years go by, the activities of living will create layers of meaning to enrich us and others.

In this way, we will be as close as we ever can be to the thing we first sought: the elusive *more*.

TRY TO BE ALIVE

In *Gatsby*, Fitzgerald suggested we accept that life is pointless and get on with it. Reject that, for it requires you to swallow a lot of darkness. Psychiatrist Viktor Frankl put it like this: "what man actually needs is not a tensionless state but rather the striving and struggling for some goal worthy of him."[140] This better path was first set down by a gifted Greek thinker three hundred years before the birth of Christ, was updated for the modern age by Dr. Frankl, and is the answer to the question written on every human heart.

Begin by identifying the virtues that matter most to you. You don't have to start from scratch, because nothing you'll come up with isn't already found in religion, philosophy, or history. Study them. Find what touches you. Then state clearly and in simple language what you choose to do: elevate your god, serve your fellow man, seek beauty, show grace. Maybe all of those. Maybe some of those. Maybe something else. Choose carefully, because you're declaring that these priorities are the foundations of a life worth living, your life.

Next, figure out what you do well and, of those things, which bring you the most joy. Choose among those activities and skills by asking how each might advance your higher aspirations. The connection doesn't have to be direct. What matters is that what you decide to do allows you to encourage yourself and the world toward those better things. In this way the dopamine pull of tomorrow is pleasantly and purposefully set in good balance with the joy of living in the here and now, and aimed overall at what you have decided is the higher purpose of your life.

Don't let that short summary trick you into thinking that reorienting your life in this way will be easy or quick. It will be neither. Do it anyway.

Reject the rudderless nihilism of the modern age. It papers over the fundamental human need for meaning. It is intellectual cant. Instead, embrace this thing you always knew was true: what the human heart longs for is real and reachable only when it is aligned with purpose—a meaning that enriches your life, and the lives of those around you, and the lives of those who will follow.

There will be suffering, but what life doesn't come with that? Purpose is not the perfect path, but it is the best of the imperfect choices left to imperfect beings. At least we get to rely on the character we've cultivated and the talent we've honed, the better to appreciate the here and now while we're in it. We can take pride in matching our self-reliance with the forever desire to return some kindness for what we've received. The writer William Saroyan, a man of profound passions, advises us to live here with intensity:

> *Try to learn to breathe deeply, really to taste food when you eat, and when you sleep really to sleep. Try as much as possible to be wholly alive with all your might, and when you laugh, laugh like hell. And when you get angry, get good and angry. Try to be alive.*[141]

THE SOUND OF A TUNING FORK, STRUCK UPON A STAR

The moment you're in? That's all you have, so choose your pursuits well. Make them pleasurable expressions of your highest aspirations, and make your aspirations an expression of whatever is good and true. Having chosen well how to fill your days, experience them fully. Work hard at this. Deserve the privilege of existence. Try to be alive.

AUTHOR'S NOTE

This book would not be as thorough as it is without the hard work of my friend Dan Lieberman, who produced first drafts of some of the more technical portions, shared copious links and research, and engaged me in the back-and-forth that rigorous thinking requires if you're going to make a complicated subject useful and appealing. He's also a psychiatrist, so he gets credit for most of the doctor stuff. I think of him as the John to my Paul, though the hair doesn't match for either of us. Dan prefers the Ramones, anyway. Thanks, friend.

That said, any oversimplifications, inaccuracies, errors, and outright falsehoods—I'm thinking, for example, about page 101 wherein I reproduce my Best Screenplay Oscar speech from 2005—are entirely my own.

Thanks to my wife, Julia, who sat alone a lot of hours while I was holed up in my office with this. She has an excellent ear, gave me smart feedback, and handed me more than a few original ideas in this book that I have claimed as my own.

Also thanks to the folks who have listened to me talk about this for years. What great friends I have! Thanks to writer, teacher, steak-dinner partner, confidant, and fellow parent-of-twins Jonathan Rick; constant phone companion Irene Schindler; fifteen years of students at Georgetown University; the always-encouraging Craig Colgan; my dear friend Salwa Emerson, who makes me feel better about everything; painter

AUTHOR'S NOTE

and deep thinker Roger Banks; the ever-faithful Jimmy Hubbard; the Sunday night writing group; my brother, the true king of jest, Todd Long; and my kids Sam (and Tabi and Leo and Phoebe), Madeline (and Alex), and Brynne (and her dog, the traitorous Benedict Arnold). Special thanks to Cody Marley and Lawrence Thomas for a lifetime. Biggest thanks of all to Mom. I love you.

Thanks, too, to my agents Wendy Levinson and Andrea Somberg of the Harvey Klinger Agency in glorious Manhattan—how do you keep track of so many things? Plus extra-special thanks to editor Leah Wilson, who (very patiently) made this book better than it was at the start, and copy editor James Fraleigh, whose artistry on both *Molecule* manuscripts makes me think we should all be working for him and not the other way around.

All praise to the BenBella team: Glenn Yeffeth, Jennifer Canzoneri, Rachel Phares, Isabelle Rubio, Madeline Grigg, and the rest of the ever-busy bunch that's been there for me throughout. I love it when I'm having a conversation and I get to say, "My publisher says . . ." I will never, ever, ever get over the fact that a country boy from a splendid wide spot in the road has been blessed with such a delightful life as a writer, teacher, and talker.

Finally, these are the writers who showed me how to write and made me want this way of living, thinking, and creating: John Irving, Tom Wolfe, James Leo Herlihy, Ira Levin, Neil LaBute, and Vince Gilligan. They write sentences, plots, and beautiful things that make me want to toss their stuff across the room and ask myself why I even try. But even then I'm smiling. Their examples provide the most joyful encouragement there is.

Novel coming soon. Dear reader, I hope you're as interested in this forthcoming flight of imagination as you have been in what I've shared so far. Thanks.

Michael E. Long
October 2024

NOTES

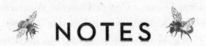

1. Woody Allen, dir., *Crimes and Misdemeanors* (1989; Orion Pictures), DVD.
2. John Hughes, dir., *Ferris Bueller's Day Off* (1986; Paramount Pictures, United International Pictures), DVD.
3. N. D. Volkow et al., "Evidence That Sleep Deprivation Downregulates Dopamine D2R in Ventral Striatum in the Human Brain," *Journal of Neuroscience* 32, no. 19 (May 9, 2012): 6711–17, https://doi.org/10.1523/JNEUROSCI.0045-12.2012.
4. K. S. Korshunov, L. J. Blakemore, and P. Q. Trombley, "Dopamine: A Modulator of Circadian Rhythms in the Central Nervous System," *Frontiers in Cellular Neuroscience* 11 (April 3, 2017): 91, https://doi.org/10.3389/fncel.2017.00091.
5. H. Chapin et al., "Dynamic Emotional and Neural Responses to Music Depend on Performance Expression and Listener Experience," *PLOS ONE* 5, no. 12 (2010): e13812; S. Koelsch, "Brain Correlates of Music-Evoked Emotions," *Nature Reviews Neuroscience* 15 (2014): 170–80; Eric W. Dolan, "Psychological and Neurobiological Foundations of Musical Pleasure Illuminated by New Research," *PsyPost*, December 7, 2023.
6. T. W. Kjaer et al., "Increased Dopamine Tone During Meditation-Induced Change of Consciousness," *Cognitive Brain Research* 13, no. 2 (April 2002): 255–59, https://doi.org/10.1016/s0926-6410(01)00106-9.
7. "Rainn Wilson—Club Random with Bill Maher," *Club Random Podcast*, YouTube video, 1:13:42, July 9, 2023, https://www.youtube.com/watch?v=nAZ9T5qcD2k.
8. A. A. Milne and E. H. Shepard, *Winnie-the-Pooh*, 1st ed. (Dutton, 1974).
9. American Association of Suicidology, "Some Facts About Suicide and Depression," June 23, 2009, https://www.cga.ct.gov/asaferconnecticut/tmy/0129/Some%20Facts%20About%20Suicide%20and%20Depression%20-%20Article.pdf
10. H. Kadry, B. Noorani, and L. Cucullo, "A Blood-Brain Barrier Overview on Structure, Function, Impairment, and Biomarkers of Integrity," *Fluids and Barriers of the CNS* 17 (2020): 69, https://doi.org/10.1186/s12987-020-00230-3.

NOTES

11 PsychonautRyan, "Stimulants—Amphetamine-Induced Narcissism," Bluelight.org, September 15, 2013, https://bluelight.org/xf/threads/amphetamine-induced-narcissism.689506/.

12 Sharlene Kaye and Shane Darke, "The Diversion and Misuse of Pharmaceutical Stimulants: What Do We Know and Why Should We Care?" *Addiction* 107, no. 3 (2012): 467–77.

13 Daniel A. Burgard et al., "Potential Trends in Attention Deficit Hyperactivity Disorder (ADHD) Drug Use on a College Campus: Wastewater Analysis of Amphetamine and Ritalinic Acid," *Science of the Total Environment* 450 (2013): 242–49.

14 Karen L. Cropsey et al., "Mixed-Amphetamine Salts Expectancies Among College Students: Is Stimulant Induced Cognitive Enhancement a Placebo Effect?" *Drug and Alcohol Dependence* 178 (2017): 302–9.

15 Cook, C., Kurtz-Costes, B., & Burnett, M. (2021). "Nonprescription Stimulant Use at a Public University: Students' Motives, Experiences, and Guilt," *Journal of Drug Issues*, 51(2), 376-390. https://doi.org/10.1177/0022042620988107.

16 Beth E. Snitz et al., "Ginkgo Biloba for Preventing Cognitive Decline in Older Adults: A Randomized Trial," *Journal of the American Medical Association* 302, no. 24 (2009): 2663–70.

17 Anna C. Nobre et al., "L-theanine, a Natural Constituent in Tea, and Its Effect on Mental State," *Asia Pacific Journal of Clinical Nutrition* 17 Suppl. 1 (2008): 167–68.

18 "2019 CRN Consumer Survey on Dietary Supplements," *Council for Responsible Nutrition*, www.crnusa.org/2019survey.

19 Regan L. Bailey et al., "Why US Adults Use Dietary Supplements," *JAMA Internal Medicine* 173, no. 5 (2013): 355–61.

20 Ghodarz Akkasheh et al., "Clinical and Metabolic Response to Probiotic Administration in Patients with Major Depressive Disorder: A Randomized, Double-Blind, Placebo-Controlled Trial," *Nutrition* 32, no. 3 (2016): 315–20.

21 Lenka Dohnalová et al., "A Microbiome-Dependent Gut-Brain Pathway Regulates Motivation for Exercise," *Nature* (2022): 1–9.

22 Qiangchuan Hou et al., "Probiotic-Directed Modulation of Gut Microbiota Is Basal Microbiome Dependent," *Gut Microbes* 12, no. 1 (2020): 1736974, https://doi.org/10.1080/19490976.2020.1736974

23 Carolina Cueva et al., "Understanding the Impact of Chia Seed Mucilage on Human Gut Microbiota by Using the Dynamic Gastrointestinal Model SIMGI®," *Journal of Functional Foods* 50 (2018): 104–11, https://doi.org/10.1016/j.jff.2018.09.028.

24 Cameron Sepah, "Dopamine Fasting 2.0—the Hot Silicon Valley Trend," LinkedIn, November 4, 2019, https://www.linkedin.com/pulse/dopamine-fasting-new-silicon-valley-trend-dr-cameron-sepah.

25 Nellie Bowles, "How to Feel Nothing Now, in Order to Feel More Later," *New York Times*, November 7, 2019, https://www.nytimes.com/2019/11/07/style/dopamine-fasting.html.

NOTES

26 Ron Gutman, *Smile: The Astonishing Power of a Simple Act* (New York: TED Books, 2011).

27 R. D. Norman, "I Hope You Have the Fortune to Experience Misfortune," YouTube video, 12:39, October 10, 2022, https://www.youtube.com/watch?v=HGFd6MbE3vA.

28 Roberts, J.G., Lugo-Morales, L.Z., Loziuk, P.L., Sombers, L.A. "Real-time chemical measurements of dopamine release in the brain," Methods Mol Biol. 2013;964:275-94. https://doi.org/10.1007/978-1-62703-251-3_16. PMID: 23296789; PMCID: PMC5259961.

29 "In Vivo Fast-Scan Cyclic Voltammetry (FSCV) for Recording Dopamine Transients with a Carbon Fiber Microelectrode," *ChemistrySelect*, accessed March 8, 2024, https://chemistry-europe.onlinelibrary.wiley.com/doi/10.1002/slct.202203630.

30 Andres M. Lozano et al., "Deep Brain Stimulation: Current Challenges and Future Directions," *Nature Reviews Neurology* 15 (2019): 148–60, https://www.nature.com/articles/s41582-018-0128-2.

31 Susan A. Reedijk, Anne Bolders, and Bernhard Hommel, "The Impact of Binaural Beats on Creativity," *Frontiers in Human Neuroscience* 7 (2013), accessed September 18, 2024, https://www.frontiersin.org/journals/human-neuroscience/articles/10.3389/fnhum.2013.00786/full.

32 "Brainwaves: The Language of Quantum Medicine?," NHA Health, accessed March 8, 2024, https://nhahealth.com/brainwaveas-the-language/.

33 Reedijk, *op. cit.*

34 "Whether physical exertion feels 'easy' or 'hard' may be due to dopamine levels, study suggests," *ScienceDaily*, April 12, 2023, https://www.sciencedaily.com/releases/2023/04/230412131051.html.

35 Adam B. Weinberger, Adam E. Green, and Evangelia G. Chrysikou, "Using Transcranial Direct Current Stimulation to Enhance Creative Cognition: Interactions Between Task, Polarity, and Stimulation Site," *Frontiers in Human Neuroscience* 11 (2017): 246.

36 "Discovery and Development of Warfarin," American Chemical Society, accessed March 8, 2024, https://www.acs.org/education/whatischemistry/landmarks/warfarin.html.

37 Douwe Draaisma, "Lithium: The Gripping History of a Psychiatric Success Story," *Nature*, August 26, 2019, https://www.nature.com/articles/d41586-019-02480-0.

38 Natalia Gulyaeva and Victor Aniol, "Good Guys from a Shady Family," *Journal of Neurochemistry* 121, no. 6 (2012): 841–42.

39 Comment on: "Is Dopamine Fasting Real?," YouTube, accessed March 8, 2024, https://www.youtube.com/watch?v=dxuM-dGJfyA.

40 Gina Kolata, "Parkinson's Research Is Set Back by Failure of Fetal Cell Implants," *New York Times*, March 8, 2001, https://www.nytimes.com/2001/03/08/us/parkinson-s-research-is-set-back-by-failure-of-fetal-cell-implants.html.

NOTES

41 "Novel Cell Therapy Continues to show Promise for Parkinson's Disease," UCI Health, March 6, 2024, https://www.ucihealth.org/news/2024/03/parkinsons-clinical-trial-18-month-results.

42 Antonio Regalado, "A Biotech Company Says It Put Dopamine-Making Cells into People's Brains," *MIT Technology Review*, September 13, 2023, https://www.technologyreview.com/2023/08/31/1078728/biotech-dopamine-making-cells-into-peoples-brains/.

43 Ibid.

44 "Miracle Drugs Like Wegovy and Ozempic Could Cure Destructive Addictions like Drinking, Gambling and Excessive Shopping by Blocking Brain's Pleasure Receptors," *Daily Mail*, April 12, 2024, https://www.dailymail.co.uk/health/article-12111143/Wegovy-Ozempic-anti-addiction-drugs-cure-drinking-shopping-habits.html.

45 Daniel Vallöf, Jesper Vestlund, and Elisabeth Jerlhag, "Glucagon-Like Peptide-1 Receptors Within the Nucleus of the Solitary Tract Regulate Alcohol-Mediated Behaviors in Rodents," *Neuropharmacology* 149, no. 1 (2019), 124–32, https://www.sciencedirect.com/science/article/pii/S0028390819300541.

46 Ibid.

47 Nicole S. Hernandez et al., "Glucagon-Like Peptide-1 Receptor Activation in the Ventral Tegmental Area Attenuates Cocaine Seeking in Rats," *Nature*, accessed May 14, 2024, https://www.nature.com/articles/s41386-018-0010-3.

48 Lisa C. Decker et al., "Treatment with a GLP-1R Agonist and a GABA-B Agonist Reduces Compulsive-Like Food and Alcohol Seeking Behaviors in Male and Female Mice," *Neuropsychopharmacology* 45, no. 4 (2020): 718–26, https://www.ncbi.nlm.nih.gov/pmc/articles/PMC6969180/; Sarah Zhang, "The Appetite-Suppressing Drug That Could Treat Addiction," *The Atlantic*, May 8, 2023, https://www.theatlantic.com/health/archive/2023/05/ozempic-addictive-behavior-drinking-smoking/674098/.

49 "CNS Effects of Wegovy (Semaglutide)," *NeuroSens*, February 28, 2023, https://neuro-sens.com/cns-effects-of-wegovy-semaglutide/.

50 Emeran A. Mayer, Kirsten Tillisch, and Arpana Gupta, "Gut/Brain Axis and the Microbiota," *Journal of Clinical Investigation* 125, no. 3 (2015): 926–38, https://www.ncbi.nlm.nih.gov/pmc/articles/PMC5797481/.

51 The Association for Behavioral and Cognitive Therapies maintains a list of books they recommend at http://www.abct.org/SHBooks/.

52 Stephanie Merritt, "Never a Problem Child," *The Guardian*, September 23, 2001, https://www.theguardian.com/books/2001/sep/23/biography.billyconnolly.

53 Dennis Greenberger and Christine A. Padesky, *Mind over Mood: Change How You Feel by Changing the Way You Think* (New York: Guilford, 2015).

54 "Acceptance and Commitment Therapy," *Psychology Today*, accessed March 8, 2024, https://www.psychologytoday.com/us/therapy-types/acceptance-and-commitment-therapy.

NOTES

55 Steven C. Hayes, "A Human Life Is Not a Problem to Be Solved," *Psychology Today*, January 30, 2009, https://www.psychologytoday.com/us/blog/get-out-your-mind/200901/human-life-is-not-problem-be-solved.
56 T. W. Moore, "What Is Logotherapy?," YouTube video, 14:23, January 12, 2022, https://www.youtube.com/watch?v=2bOFLS39Mmk.
57 Traci Pedersen, "Flooding Therapy: What It Is and How It Works," *Psych Central*, August 9, 2022, https://psychcentral.com/blog/ocd-and-flooding-exposure.
58 B. Riley et al., "Graded Exposure Therapy for Online Mobile Smartphone Sports Betting Addiction: A Case Series Report," *Journal of Gambling Studies* 37, no. 4 (2021): 1263 75.
59 "The Power of the Placebo Effect," *Harvard Health*, December 13, 2021, https://www.health.harvard.edu/mental-health/the-power-of-the-placebo-effect.
60 K. G. Seshadri, "The Neuroendocrinology of Love," *Indian Journal of Endocrinology and Metabolism* 20, no. 4 (July–August 2016): 558–63, https://doi.org/10.4103/2230-8210.183479.
61 W. R. Jankowiak and E. F. Fischer, "A Cross-Cultural Perspective on Romantic Love," *Ethnology* 31, no. 2 (1992): 149–55, https://doi.org/10.2307/3773618.
62 Chase Winter, "Does the West Have a Monopoly on Romantic Love?" *The World from PRX*, February 13, 2014, https://theworld.org/stories/2014-02-12/does-west-have-monopoly-romantic-love.
63 Florian Zsok et al., "What Kind of Love Is Love at First Sight? An Empirical Investigation," *Personal Relationships* 24, no. 4 (2017): 869–85, https://doi.org/10.1111/pere.12218.
64 Arthur Aron et al., "The Experimental Generation of Interpersonal Closeness: A Procedure and Some Preliminary Findings," *Personality and Social Psychology Bulletin*, 23, no. 4 (1997): 363–77, https://doi.org/10.1177/0146167297234003.
65 Jones, Daniel, "The 36 Questions That Lead to Love," *New York Times*, January 9, 2015, accessed October 4, 2024, https://www.nytimes.com/2015/01/09/style/no-37-big-wedding-or-small.html.
66 J. W. Pennebaker et al., "Don't the Girls' Get Prettier at Closing Time: A Country and Western Application to Psychology," *Personality and Social Psychology Bulletin* 5, no. 1 (1979): 122–25, https://doi.org/10.1177/014616727900500127.
67 Ibid.
68 University of Western Australia, "Men who Take Risks Are More Attractive to Women, Study Suggests," *Neuroscience News*, February 27, 2023, https://neurosciencenews.com/attraction-risk-taking-men-22560/.
69 Cyril C. Grueter et al., "Preference for Male Risk Takers Varies with Relationship Context and Health Status but Not COVID Risk," *Evolutionary Psychological Science* 9, no. 3 (2023): 283–92, https://doi.org/10.1007/s40806-023-00354-3.
70 Ibid.
71 Ibid.
72 Patrick McAlvanah, "Are People More Risk-Taking in the Presence of the Opposite Sex?" *Journal of Economic Psychology* 30, no. 2 (2009): 136–46; B.

NOTES

Pawlowski, R. Atwal, and R. I. M. Dunbar, "Sex Differences in Everyday Risk-Taking Behavior in Humans," *Evolutionary Psychology* 6, no. 1 (2008): https://doi.org/10.1177/147470490800600104.
73 Grueter, op. cit.
74 Bernhard Fink, Nick Neave, and Hanna Seydel, "Male Facial Appearance Signals Physical Strength to Women," *American Journal of Human Biology* 19, no. 1 (2007): 82–87, https://doi.org/10.1002/ajhb.20583.
75 Robin Nixon Pompa and Patrick Pester, "How to Tell You're in Love with Someone, According to the Science," *LiveScience*, February 13, 2015, https://www.livescience.com/33720-13-scientifically-proven-signs-love.html.
76 Ibid.
77 Mona Chalabi, "Dear Mona, I Masturbate More Than Once a Day. Am I Normal?" *FiveThirtyEight*, January 7, 2015, https://fivethirtyeight.com/features/dear-mona-i-masturbate-more-than-once-a-day-am-i-normal/.
78 Gregg Levoy, "Frustration Attraction—How Separation Heightens Passion," *Psychology Today*, October 26, 2020, https://www.psychologytoday.com/us/blog/passion/202010/frustration-attraction-how-separation-heightens-passion.
79 Donna J. Bridge and Ken A. Paller, "Neural Correlates of Reactivation and Retrieval-Induced Distortion," *Journal of Neuroscience* 32, no. 35 (August 29, 2012): 12144-51, https://doi.org/10.1523/JNEUROSCI.1378-12.2012.
80 Hongbo Song et al., "Improving Relationships by Elevating Positive Illusion and the Underlying Psychological and Neural Mechanisms," *Frontiers in Human Neuroscience* 12 (2019): 526, https://doi.org/10.3389/fnhum.2018.00526.
81 Shelley E. Taylor and Jonathon D. Brown, "Positive Illusions and Well-Being Revisited: Separating Fact from Fiction," *Psychological Bulletin* 116, no. 1 (1994): 21–27, https://doi.org/10.1037/0033-2909.116.1.21.
82 Woody Allen, dir., *Annie Hall* (1977; United Artists), DVD.
83 Levoy, op. cit.
84 "Thomas Sullivan," Carnegie Mellon University College of Engineering, accessed March 8, 2024, https://engineering.cmu.edu/directory/bios/sullivan-thomas.html.
85 "Multitasking: Switching Costs," American Psychological Association, accessed March 4, 2024, https://www.apa.org/topics/research/multitasking.
86 Tony McCaffrey, "Why We Can't See What's Right in Front of Us," *Harvard Business Review*, May 10, 2012, https://hbr.org/2012/05/overcoming-functional-fixednes.
87 Jackson G. Lu, Modupe Akinola, and Malia F. Mason, "'Switching On' Creativity: Task Switching can Increase Creativity by Reducing Cognitive Fixation," *Organizational Behavior and Human Decision Processes* 139 (2017): 63–75.
88 Shayla Cunningham, Corey Hudson, and Kate Harkness, "Social Media and Depression Symptoms: A Meta-Analysis," *Journal of Abnormal Child Psychology* 49, no. 2 (2021): 241–53.
89 Matt Labash, "Be This Guy, Instead of the Angry Jerk You're Becoming," Substack, January 15, 2024, used by permission.

NOTES

90 Dane Mauer-Vakil and Anees Bahji, "The Addictive Nature of Compulsive Sexual Behaviours and Problematic Online Pornography Consumption: A Review," *Canadian Journal of Addiction* 11, no. 3 (September 2020): 42–51.
91 Claudio Vieira and Mark D. Griffiths, "Problematic Pornography Use and Mental Health: A Systematic Review," *Sexual Health & Compulsivity 31*, no. 3 (2024): 207–247.
92 Joshua B. Grubbs, Shane W. Kraus, and Samuel L. Perry, "Self-reported Addiction to Pornography in a Nationally Representative Sample: The Roles of Use Habits, Religiousness, and Moral Incongruence," *Journal of Behavioral Addictions* 8, no. 1 (2019): 88–93.
93 Meghan Donevan, Marie Bladh, Åsa Landberg, L.S. Jonsson, Gisela Priebe, Inga Dennhag, and Carl Goran Svedin, "Closing the Gender Gap? A Cohort Comparison of Adolescent Responses to and Attitudes Toward Pornography, 2004 vs. 2021," *The Journal of Sex Research* (October 2024): 1–15.
94 Sean McNabney, Krisztina Hevesi, and David L. Rowland, "Effects of Pornography Use and Demographic Parameters on Sexual Response During Masturbation and Partnered Sex in Women," *International Journal of Environmental Research and Public Health* 17, no. 9 (2020): 3130.
95 Daniel A. Cox, Beatrice Lee, and Dana Popky, "Politics, Sex, and Sexuality: The Growing Gender Divide in American Life," accessed September 20, 2024, https://www.americansurveycenter.org/research/march-2022-aps/.
96 David Shultz, "Divorce Rates Double When People Start Watching Porn," *Science*, August 26, 2016.
97 Thibault Jacobs et al., "Associations Between Online Pornography Consumption and Sexual Dysfunction in Young Men: Multivariate Analysis Based on an International Web-Based Survey," *JMIR Public Health and Surveillance* 7, no. 10 (2021): e32542, https://doi.org/10.2196/32542.
98 Faraz Ahmed, M. Zubair Shafiq, and Alex X. Liu, "The Internet is For Porn: Measurement and Analysis of Online Adult Traffic," 2016 IEEE 36th International Conference on Distributed Computing Systems (2016): 88, https://doi.org/10.1109/ICDCS.2016.81.
99 Mark Regnerus, David Gordon, and Joseph Price, "Documenting Pornography Use in America: A Comparative Analysis of Methodological Approaches," *The Journal of Sex Research* 53, no. 7 (2016): 873–81, https://doi.org/10.1080/00224499.2015.1096886.
100 Ceci, Laura, "Adult and pornographic website industry market size in the U.S. 2018-2023," Statista, accessed October 5, 2024, https://www.statista.com/statistics/1371582/value-online-website-porn-market-us/#statisticContainer.
101 Azcuna, Lyndon, "The Porn Pandemic," LifePlan, October 28, 2021, accessed October 5, 2024, https://www.lifeplan.org/the-porn-pandemic/#:~:text=The%20porn%20industry%20generates%20more,is%20being%20spent%20on%20pornography.
102 "Total revenue of all National Football League (NFL) teams from 2001 to 2023," Statista, accessed October 5, 2024, https://www.statista.com/statistics/193457/

total-league-revenue-of-the-nfl-since-2005/#:~:text=Total%20revenue%20of%20 the%20NFL%202001%2D2023&text=In%202023%2C%20the%2032%20 teams,dollars%20over%20the%20previous%20year.

103 Joshua B. Grubbs, Shane W. Kraus, and Samuel Perry, "Self-Reported Addiction to Pornography in a Nationally Representative Sample: The Roles of Use Habits, Religiousness, and Moral Incongruence," *Journal of Behavioral Addictions* 8, no. 1 (2019): 88–93.

104 "Gluten Intolerance," Cleveland Clinic, accessed October 5, 2024, https:// my.clevelandclinic.org/health/diseases/21622-gluten-intolerance.

105 "The Top 10 Leading Causes of Death in the United States," Healthline, accessed March 8, 2024, https://www.healthline.com/health/leading-causes-of-death.

106 Luke Sniewski et al., "Meditation as an Intervention for Men with Self-Perceived Problematic Pornography Use: A Series of Single Case Studies," *Current Psychology* 41 (2022): 5151–62, https://doi.org/10.1007/s12144-020-01035-1.

107 Dean P. Fernandez, Daria J. Kuss, and Mark D. Griffiths, "The Pornography 'Rebooting' Experience: A Qualitative Analysis of Abstinence Journals on an Online Pornography Abstinence Forum," *Archives of Sexual Behavior* 50, no. 2 (2021): 711–28, https://doi.org/10.1007/s10508-020-01858-w.

108 S. S. Alavi et al., "Behavioral Addiction Versus Substance Addiction: Correspondence of Psychiatric and Psychological Views," *International Journal of Preventive Medicine* 3, no. 4 (2012): 290–94.

109 U.S. Census Bureau, "Real Median Household Income in the United States [MEHOINUSA672N]," retrieved from FRED, Federal Reserve Bank of St. Louis, accessed September 27, 2024, https://fred.stlouisfed.org/series/ MEHOINUSA672N.

110 "Working Hours in the US—Overview," Clockify, accessed March 8, 2024, https:// clockify.me/working-hours.

111 "U.S. GDP 1960–2023," Macrotrends, accessed March 8, 2024, https:// www.macrotrends.net/countries/USA/united-states/gdp-gross-domestic-product#:~:text=U.S.%20gdp%20for%202022%20was,a%204.13%25%20 increase%20from%202018.

112 Mary Kekatos, "US Life Expectancy Falls to Lowest Levels Since 1996 due to COVID-19, Drug Overdoses," *ABC News*, August 31, 2022, https://abcnews. go.com/Health/us-life-expectancy-falls-lowest-levels-1996-due/story?id=95649464.

113 The Ness Center. "Unraveling the Link Between Compulsive Buying and Mental Health," accessed October 3, 2024, https://thenesscenter.com/unraveling-the-link-between-compulsive-buying-and-mental-health/.

114 Donald W. Black, "A Review of Compulsive Buying Disorder," *World Psychiatry* 6, no. 1 (2007): 14–8.

115 Substance Abuse and Mental Health Services Administration, "Table 3.13: DSM-IV to DSM-5 Obsessive-Compulsive Disorder Comparison," in *Impact of the DSM-IV to DSM-5 Changes on the National Survey on Drug Use and Health* (Rockville, MD: Author, June 2016), https://www.ncbi.nlm.nih.gov/books/NBK519704/table/ch3.t13/.

NOTES

116 Jillian Seiler et al., "Dopamine Signaling in the Dorsomedial Striatum Promotes Compulsive Behavior," *Current Biology* 32, no. 5 (2022): 1175–88, https://www.cell.com/current-biology/fulltext/S0960-9822(22)00117-8.
117 "Game Industry Usage and Revenue Statistics 2023," Helplama.com, January 8, 2024.
118 Tess Farrand, "Who Goes to the Movies?" Movieguide, November 16, 2018.
119 Christina Gough, "Topic: Sports on TV," Statista, March 1, 2024.
120 "Videogames Are a Bigger Industry Than Movies and North American Sports Combined, New Report Finds," MarketWatch, accessed March 9, 2024.
121 Bedbible Research Center, "Porn Industry Revenue Numbers & Stats," Bedbible.com, August 23, 2023.
122 Raelene Knowles, "The Power of Play," IGEA, October 9, 2023.
123 J. Clement, "U.S. Single Player vs. Multiplayer Frequency among Gamers 2022," Statista, January 26, 2024.
124 Private correspondence with Drake Greer, March 14–16, 2024.
125 "Christoph Niemann: Illustration," *Abstract: The Art of Design*, created by Scott Dadich, Netflix, February 10, 2017.
126 Attributed variously to comedian Jack Benny, violinist Jascha Heifetz, pianist Arthur Rubinstein, and writer and publisher Bennett Cerf; this speculation courtesy of Michael Pollak, "The Origins of That Famous Carnegie Hall Joke," *New York Times*, November 27, 2009, https://www.nytimes.com/2009/11/29/nyregion/29fyi.html.
127 Roger E. Beaty et al., "Robust Prediction of Individual Creative Ability from Brain Functional Connectivity," *Proceedings of the National Academy of Sciences* 115, no. 5 (2018): 1087–92.
128 Todd I. Lubart, "Models of the Creative Process: Past, Present and Future," *Creativity Research Journal* 13, nos. 3–4 (2001): 295–308.
129 Donghwy An and Nara Youn, "The Inspirational Power of Arts on Creativity," *Journal of Business Research* 85 (2018): 467–75.
130 Chin-Wen Yeh, Shih-Han Hung, and Chun-Yen Chang, "The Influence of Natural Environments on Creativity," *Frontiers in Psychiatry* 13 (2022): 895213, https://doi.org/10.3389/fpsyt.2022.895213.
131 Rachel Kaplan and Stephen Kaplan, *The Experience of Nature: A Psychological Perspective* (Cambridge University Press, 1989).
132 Janetta Mitchell McCoy and Gary W. Evans, "The Potential Role of the Physical Environment in Fostering Creativity," *Creativity Research Journal* 14, nos. 3–4 (2002): 409–26.
133 Sylvie Studente, Nina Seppala, and Noemi Sadowska, "Facilitating Creative Thinking in the Classroom: Investigating the Effects of Plants and the Colour Green on Visual and Verbal Creativity," *Thinking Skills and Creativity* 19 (2016): 1–8.
134 Célia Lacaux et al., "Sleep Onset Is a Creative Sweet Spot," *Science Advances* 7, no. 50 (2021): eabj5866, https://doi.org/10.1126/sciadv.abj5866.
135 *John Mulaney: The Comeback Kid* (2015, Netflix).

NOTES

136 E. P. Torrance, *The Torrance Tests of Creative Thinking Norms—Technical Manual Figural (Streamlined)* (Scholastic Testing Service, Inc., 1990).
137 Arthur C. Clarke, *Profiles of the Future* (Gollancz, 1962).
138 Arthur Cropley, "In Praise of Convergent Thinking," *Creativity Research Journal* 18, no. 3 (2006): 391–404.
139 Henri Nouwen, *Life of the Beloved* (Crossroad Publishing Company, 1992).
140 Viktor E. Frankl, *Man's Search for Meaning* (Beacon Press, 2006).
141 William Saroyan, "Preface to the First Edition," *The Daring Young Man on the Flying Trapeze and Other Stories* (Random House, 1934), 12–13.

INDEX

A

abstinence
    from added sugar, 55, 60, 61
    from alcohol, 59, 188
    in dopamine revitalization/fasting, 59–60
    from drugs, 188
    from gaming, 204
    from news consumption, 175–179
    from online pornography, 188–189
    from social media, 166–170
abstractions, manipulating
    in art and mathematics, 158
    avoiding, during dopamine fasting, 57
    creativity and, 219
    dopamine and, 24, 26
    on social media, 172–173
    value identification and, 90
abstractness of titles, 219
acceptance, 87–88, 146
acceptance and commitment therapy (ACT), 84–91
    addressing addictive behaviors with, 193
    CBT and behavioral therapy vs., 84–86
    choosing to use, 98, 99
    dealing with trauma in, 187
    logotherapy and, 91
    six skills in, 87–91, 206
    to tame lust, 110
    treating dopamine-driven problems with, 86–87
accountability partner(s)
    for gaming, 202
    for pornography use, 186–187
    for shopping, 197
    for social media use, 167
    therapist as, 84
achievement
    ideal experience of, 32–33
    in meaningful/purposeful living, 230–231, 234
    overestimating satisfaction from, 33
    pleasure from, 16–17
ACT. *see* acceptance and commitment therapy
adaptation, 14, 55, 60
Adderall, 39, 43–45
addiction
    boosting normal dopamine levels and, 40
    problematic pornography use and, 181–182
    semaglutide to "cure," 77–80
    tolerance in, 30, 31
    viewing shopping as, 191–193
addictive and compulsive behaviors
    cue exposure therapy to treat, 94–97
    pornography use, 181–182
    sexual, 128–131
    shopping, 191–196
    social media use, 164–165
    therapy and pharmaceuticals to address, 193
ADHD. *see* attention deficit/hyperactivity disorder
Adler, Alfred, 91, 92, 97
advertising
    compulsive shopping encouraged by, 193
    and online pornography, 184

INDEX

on social media, 164, 165
    in video games, 200
age of exposure, to pornography, 182
aggression, 38
agonists, 79
alcohol
    abstinence from, 59
    cravings for, 30, 79
    elimination of, 60
    problematic pornography use and addiction to, 181–182
    tolerance for, 28
algorithm, social media, 163, 167–168
alpha waves, 67
alternative options, considering
    to avoid news, 178
    to control PPU, 186
    in divergent thinking, 151
    for gamers, 201, 204–206
    to reduce shopping, 197
    in romantic love, 110, 112
    to treat dopamine-related problems, 104
Alzheimer's disease, 183n*
Amazon, 72, 191
ambition, 149–150, 229, 233
American Enterprise Institute, 182
amphetamine (Adderall), 39, 43–45
amusement, 3, 168, 178
Android operating system, 166
animal studies
    of exercise motivation, 49–51, 69
    of GLP-1 agonists, 79
    of modified beta-carbolines, 74–75
    of neuron transplantation, 76
Annie Hall (film), 141
anti-anxiety drugs, 193
anticipation
    in compulsive buying disorder, 195
    and desire dopamine, 23
    dopamine level and, 70
    failing to find pleasure in, 32
    in long-term relationships, 142–143
    of sexual behavior, 132–133
antidepressants, 42–43, 73, 193
anxiety
    CBT for treatment of, 83
    dopamine levels and, 38n*, 40, 48
    modified beta-carbolines and, 75
    problematic pornography use and, 182
aptitude, creativity and, 220
Aristotle, 178, 232, 234–236
art, 158–159, 214

artemisinin, 77
artificial intelligence, 167n*, 200
aspirin, 77
Atkinson, Rowan, 158
attachment
    in "me-first" marriage, 138, 139
    outside of marriage, 136
    to the "right" one, 124
    in romantic love, 107
    separateness to maintain, 142–143
    taming, 111–112
attention, 63, 150, 151, 214
attention deficit/hyperactivity disorder (ADHD), 39, 43, 152, 214
attraction
    and closing time effect, 118–120
    in gaming, 200
    love at first sight and, 115–116
    outside of marriage, 136
    to the "right" one, 126
    to risk-taking men, 121–124
    in romantic love, 107
    taming, 110–111
autonomy, 119–121
Avenue Q (musical), 171

B

Bachelor of Engineering Studies and Arts degree, 158
back-channel communications, 133
Back to Methuselah (Shaw), 215n*
balance, 31–33, 205
BBB. *see* blood–brain barrier
BDNF (brain-derived neurotropic factor), 9
beautiful lie, 139–143, 201
Bee Gees, 67
behavioral therapy, 84, 86
beta-carbolines, 73–75
beta waves, 67
"Be This Guy, Instead of the Angry Jerk You're Becoming" (Labash), 170
Bifidobacterium bifidum, 48
binaural beats, 67–69
Biographical Inventory of Creative Behaviors, 219
biological imperatives
    overcoming, 105–106
    reproduction, 105, 108, 122
    sexual behavior vs. other urges driven by, 128, 133–134
    survival, 106, 108
biologics, 41n*

INDEX

bipolar disorder, 83
birth control pill, 183–184
Black, Donald, 195
blame, 86
blood–brain barrier (BBB), 41, 42, 46
blood glucose levels, 10, 77
blood tests for dopamine, 65
body temperature, maintaining, 54
Boomers, 171
boredom, 3, 21, 31, 230
Bowles, Nellie, 56–57
box breathing, 95
brain. *see also specific structures by name*
    building connections in, 211–213, 222
    carbon quantum dots in research on, 66
    compulsions and structures of, 195–196
    direct addition of dopamine to, 40–42
    dopamine subsystem in, 23–24
    effect of binaural beats on, 68
    encoding of craving in, 30
    evolution of, to ensure survival, 18
    neuron transplantation in, 75–76
    overclocking the, 39–40, 71
brain-derived neurotropic factor (BDNF), 9
brainstorming, 68, 153–154
brainwaves, 67–68
Breaking Bad (TV series), 213
Brown, Jonathon, 141
bulimia, 83
buprenorphine, 30
B vitamins, 48

C

Cade, John, 73
calcium, 46
cancer, 183n*
carbon quantum dots (CQDs), 66
Carnegie Mellon University, 158
Carter, Jimmy, and administration, 194
Castaway (film), 55
CBT. *see* cognitive behavioral therapy
cell phones
    gaming on, 200, 205n*
    limiting access to, 165–166
    news consumption on, 175
    removing ability to pay with, 196
    settings related to social media on, 166, 168
challenge(s)
    dopamine revitalization as, 62–63
    in gaming, 199–200
    in meaningful or purposeful living, 233–234
Chaucer, Geoffrey, 3
chia seeds, 50
choice
    benefits of, 193–194
    to do good, 232–234
    in meaningful/purposeful living, 228–229
    of primary virtue, 236–238
cholera, 77
cigarettes, cravings for, 79
circadian rhythms, 8
clickbait, 169–170
closeness, 116–118
closing time effect, 118–121
cocaine, 30, 79, 166
cognitive behavioral therapy (CBT), 82–83
    ACT vs., 84–86
    addressing addictive behaviors with, 193
    dealing with dopamine-related challenges with, 99
    finding cause of behavior in, 197–198
cognitive defusion, 88, 89, 96, 112. *see also* observing self
cognitive fatigue, 154
cognitive fixation, 160–161
cohabitation, 136n*
comments, posting, 174
commitment
    and psychological flexibility, 90–91
    self-denial as demonstration of, 137–138
    to taming dopamine, 104
    to therapy, 81–82
communication, to build anticipation for sex, 132–133
competition, 127–128, 154
compulsions, 195–196. *see also* addictive and compulsive behaviors
compulsive buying disorder, 194–196
compulsive shopping (oniomania), 79, 193
concentration, 43, 44, 214
confidence, 70–71, 124
conflict
    online, 171–174
    between wants and needs, 14
connection
    and attachment, 111
    discussions to build, 117
    between ideas, 211
    moderating gaming with, 201–202
    self-denial to sustain, 132–133
    on social media, 172

INDEX

connections
    neuronal, 211–213, 222
Connolly, Billy, 84
consequences, of sexual behavior, 131
consistency, of attachment, 111
consumerism, 193–194
control dopamine circuit, 22, 24
    in ADHD, 43, 152
    in attraction, 107
    in convergent thinking, 151
    creativity and, 72, 212
    in dating "dangerous" men, 122
    during gaming, 199
    influence of, in problem behaviors, 109
    mania and, 32
    for mental time travel, 24–26
    preventing tolerance with, 29
    in search for the "right" one, 113–114, 126
convergent thinking
    for art appreciation, 159
    building capacity for, 155–157
    creativity and, 210
    defined, 151–152
coping strategies, 95, 191
Coprococcus eutactus, 49, 50
Costco effect, 142–143
courage, 236
COVID-19, 183n*
CQDs (carbon quantum dots), 66
Cranston, Bryan, 213
cravings
    in addictive/compulsive behaviors, 129
    for dopaminergic stimulation, 61
    and downregulation, 30
    nucleus accumbens and, 80
    in problematic pornography use, 181
    semaglutide and, 77–80
    for sugar, 60
creative potential, 218–220
Creative Product Inventory, 219
Creative Product Semantic Scale, 219
creativity
    binaural beats and, 68–69
    building neural connections for, 211–213
    dreaming and, 214–218
    increasing, 209–222
    inspiration for, 213–214
    in magic trick design, 209–210
    mind wandering and, 210–211
    new ways of thinking to increase, 220–222
    tests of creative potential, 218–220
    transcranial stimulation to increase, 71–73
credit cards, 196–197
Crimes and Misdemeanors (film), 4
cue exposure therapy, 94–97
culture, 106, 173
curiosity, 18, 32, 125, 137
customer returns, 154–155

D

daily living
    dopamine cycle in, 225–226
    dopamine fasting as part of, 59
    meaningful/purposeful, 227–239
Dalí, Salvador, 216
"dangerous" men, appeal of, 121–124
Darwin, Charles, 211
data harvesting, by social media, 167–168
"date," with sexual partner, 132–133
daydreaming, 143, 212
decision making
    about finding the "right" one, 121, 124
    about sexual behavior, 134
    dopamine subsystems in, 24–26
Declaration of Independence, 235n*
deep conversation, 117, 118
default mode network, 212
delta waves, 67
delusions, 38
De Niro, Robert, 213
depression
    dopamine level and activity in, 31, 37, 40
    problematic pornography use and, 182
    problematic social media use and, 164–165
    treatments for, 42, 42–43, 83
desire dopamine circuit, 22–24
    in ADHD, 43, 152
    in attraction, 107
    creativity and, 212
    depression and, 31
    in divergent thinking, 151
    effects of dopamine release in, 80
    during gaming, 199
    influence of, in problem behaviors, 109
    mania and, 32
    for mental time travel, 24–26
    in search for the "right" person, 113, 125–126
despair, 16–17, 31
diabetes, 40, 50, 77–78, 99, 183n*

INDEX

dialectical behavior therapy, 193
Dietary Supplement Health and Education Act, 45–46
discipline
    for dopamine revitalization, 63
    and happiness vs. survival, 15–16
    and need to seek therapy, 99–100
    to pursue a meaningful/purposeful life, 228–229
    for self-directed therapy, 82
    to tame the power of dopamine, 104
discomfort, 82, 85–87
dissatisfaction
    after dopamine revitalization, 62
    dopamine as cause of, 19–20
    as motivator, 16
    success and, 16–17 (*see also* dopamine cycle [dopamine chase])
dissonance theory, 120, 121
distraction, 103
    problem behaviors as, 110
    sex as, 131
    to tame lust, 109–110, 112
divergent thinking
    for art appreciation, 159
    building capacity for, 153–155
    convergent thinking as complement to, 151–152
    creativity and, 210, 213, 214
    defined, 150–151
    encouraging, 221–222
    in Torrance Test of Creative Thinking, 218–219
    while falling asleep, 217
divorce, 182
documenting
    aspirations, 237
    creative thoughts, 210, 211
    ideas upon waking up, 217–218
    pornography use, 187–188
    reasons to stop shopping, 197–198
    time spent gaming, 202–203
    your own attractive qualities, 110–112
doing good, 232–234
"Don't the Girls All Get Prettier at Closing Time" (song), 119
L-dopa (levodopa), 42, 46
dopamine
    in bloodstream, 41n^2
    cautious use of, 103–104
    direct addition of, to brain, 40–42
    dual nature of, xiv

    elimination of, 53
    function of, 19–20
    H&N neurotransmitters and, 20–21, 31–33
    influence of, on behavior, 1–2, 13–14
    and promise of pleasure, 236–237
dopamine activity
    attempting to boost normal, 39–51
    binaural beats to alter, 67–69
    building skill of regulating, 4, 13
    driving, 7–10
    lowering, with dopamine fasting, 53–63
    low vs. elevated, 37–38
    measuring levels of, 65–66
    and motivation to exercise, 69–71
    promising research on, 65–80
    seeking balance in, 31–33
    during sleep, 215
    therapy to change, 81–100
dopamine agonists, 40
Dopamine Brain Food, 46
dopamine cycle (dopamine chase)
    in daily living, 225–226
    despair due to, 16–17
    downregulation and, 29–30
    enjoyment of life and, 61
    Viktor Frankl on, 92
    news consumption and, 177–178
    in online debates, 173–174
    pursuit of meaning/purpose and, 229–230, 233–235
    shopping and, 195
    social media use and, 163–164
"dopamine drip," in gaming, 199–200
dopamine-driven problems
    "diagnosing," 99–100
    with gaming, 199–207
    increasing creative capacity, 209–222
    lion tamer mindset about, 103–104
    in long-term romantic relationships, 135–147
    with news consumption, 175–179
    with online pornography use, 181–189
    overlapping solutions to, 104
    with productivity, 149–161
    with romantic love, 105–112
    searching for the "right" person, 113–126
    selecting therapeutic approach to deal with, 97–99
    semaglutide and other GLP-1 agonists to reduce, 77–80
    with sexual behavior, 127–134

INDEX

with shopping behavior, 191–198
with social media use, 163–174
dopamine fasting. *see* dopamine revitalization
dopamine reuptake inhibitors (DRIs), 41–42
dopamine revitalization (dopamine fasting), 53–63
    as challenge, 62–63
    defined, 53–54
    enjoyment of life with, 61–62
    foundational approach to, 58–61
    homeostatic mechanisms in, 54–56
    mistakes in, 56–58
dopamine system. *see also* control dopamine circuit; desire dopamine circuit
    avoiding empty stimulation of, 168–170
    circadian rhythms and, 8
    conflict of wants and needs due to, 14
    evolution of, 2–3, 14, 18, 63
    exercise habit and, 7
    GLP-1 agonists and, 79–80
    during good surprises, 26–28
    impact of transcranial stimulation on, 71–73
    mental time travel and, 24–26
    in modern world, 18
    modified beta-carbolines and, 73–75
    music and, 8
    side effects of altering, 74
    social media's hijacking of, 163–165
    speed of responses by, 58
    stimulation and tolerance in, 28–31
    transplanting neurons involved in, 75–77
DopaRush Cocktail, 46
dorsolateral striatum, 195
dorsomedial striatum, 195
downregulation
    with Adderall use, 44
    of dopamine, 29–30, 55
    homeostatic, 55, 61
    with online pornography use, 182
dreaming, 214–218
DRIs (dopamine reuptake inhibitors), 41–42
drugs. *see also* pharmaceuticals
    abstinence from, 60
    cocaine, 30, 79, 166
    creativity and use of, 215
    limiting access to, 166
    problematic pornography use vs. addiction to, 181–182
    tolerance to, 29n*, 30
Duolingo, 159
Dylan, Bob, 213

E

EBR (eye-blink rate), 66, 67
ECT (electroconvulsive therapy), 71, 72
ED (erectile dysfunction), 182, 189
Edison, Thomas, 216
effort, in creative endeavors, 221
Eisenhower matrix, 155–156
elaboration of ideas, 219
electroconvulsive therapy (ECT), 71, 72
email, communicating via, 133
Emerson, Ralph Waldo, 149
emotional well-being, 141
endorphin, 1, 57, 99, 182
energy, 38, 43
enjoyment, dopamine revitalization and, 61–63
Entertainment Software Association, 199
enthusiasm, 23
Epicurus, 63
erectile dysfunction (ED), 182, 189
estrogens, 107
Eubacterium rectale, 49, 50
euphoria, 43
executive functioning, 9, 43, 44
exenatide, 79
exercise and physical activity, 7
    "addiction" to, 129
    as alternative to problematic behavior, 186, 205
    to deal with heartbreak, 146–147
    as habit, 186, 200
    motivation to, 49–51, 69–71
exploration, 213
exposure therapy, 93–97
external validation, of creative potential, 218–220
eye, homeostatic mechanisms of, 54
eye-blink rate (EBR), 66, 67

F

Facebook, 165, 166, 170, 172
facial masculinity, 123
failure, 63, 204
family therapy, 193
fascination, 214
fatty acid amines, 49
FDA. *see* Food and Drug Administration
fear, overcoming, 93–94
feedback, for social media algorithm, 167–168
feelings
    acceptance of, 87–88
    from attachment, 111
    cognitive defusion for, 88

INDEX

"diagnosing" situations involving, 99–100
    at end of romantic relationships, 146–147
"fencing," in intimate sessions, 133
fermented foods, 50
Fetterman, John, 165
fiber, 50
fight-or-flight system, 94
final exams, performance on, 44–45
financial counseling, 193
first meeting, with romantic partner, 113–116
Fische, Ted, 106
Fischer, Helen, 107, 136
Fitzgerald, F. Scott, 227–230, 237
flooding (exposure therapy), 93–94
fluency of ideas, 218
focus, 43, 44
Food and Drug Administration (FDA), 74, 76, 183
foods, "dopamine-enhancing," 9–10
The 40-Year-Old Virgin (film), 105
Frankl, Viktor, 91–93, 237
French Riviera, 229
frequency, 67–68
Freud, Sigmund, 91, 92, 97
friendships
    face-to-face interactions in, 171
    between gamers, 201–202
    impact of news consumption on, 177–178
frontal lobe, 24
fulfillment, 33, 229–231, 231, 236
future(s)
    creativity and speculation about, 221
    hope for, 228
    imaging alternative, 24–26

G

GABA, 38n*, 72
gambling problems, 96–97, 104, 196
gaming, 199–207
    breaking habit of, 204–207
    building human connection and, 201–202
    cue exposure therapy to reduce, 94–95
    dopamine drip in, 199–200
    reducing time spent, 202–204
    strategies for reducing problematic, 200–201
gamma waves, 68
Generation Z, 202
genetic variability, 152
genome, 49n*
Gilbert, W. S., 115
Gilley, Mickey, 118–120

Gilligan, Vince, 213
Ginkgo biloba, 46
Giurgea, Corneliu E., 40n*
glucagon-like peptide-1 (GLP-1) agonists, 79–80
glutamate, 38n*
goal-oriented behavior, 195–196, 234–235
goal setting, 60, 118, 203
Goodwill, 191
gradual exposure, 93, 94
The Great Gatsby (Fitzgerald), 227–230, 237
Green, Paul E., 76
Greer, Drake, 201, 203–205
Grueter, Cyril, 122–123
gut biome (microbiome), 47–51, 69–71

H

habits
    compulsions and, 195–196
    dopamine sensitivity and, 58–61
    exercise, 186, 200
    gaming, 204–207
    shopping, 195
hallucinations, 38
happiness, 5
    after dopamine revitalization, 62
    choosing between survival and, 15–16
    in meaningful/purposeful living, 231, 232, 235
hard fascination, 214
hard stop, for gaming, 203
having, wanting vs., 20–22
Hayes, Steven C., 84, 87
heart disease, 50, 99, 183n*
Hebb, Donald, 212
hedonics, 219n*
Helen of Troy, 127
helping others, 147
Henchcliffe, Claire, 75–76
here-and-now, finding pleasure in, 32, 57–58
    creativity and, 221–222
    during dopamine revitalization, 61–62
    living in the present and, 90
    in long-term romantic relationships, 145
    semaglutide and, 78
    during sex, 131
here & now (H&N) neurotransmitters
    balancing dopamine and, 31–33
    being in love and, 124
    during dopamine revitalization, 57–59
    at end of relationship, 146
    interactions of dopamine and, 20–21

INDEX

during sleep, 215
Hewlett-Packard, 57
high blood pressure, 50, 77
H&N neurotransmitters. *see* here & now neurotransmitters
homeostasis, 54–56, 60, 61
honesty, 117, 118, 146
hope, 93, 109, 227–229
"How to Feel Nothing Now, in Order to Feel More Later" (Bowles), 56–57
"How to Live on Practically Nothing a Year" (Fitzgerald), 229
Hutton, James, 211
Huxley, Aldous, 214
hypnagogia, 216–218
hypothyroidism, 77

I

"I Hope You Have the Fortune to Experience Misfortune" (Norman), 62–63
"imaginary camera" exercise, 145
importance
  on Eisenhower matrix, 156–157
  of news, 179
improvisation, 212
impulse control, 38, 43
incubation effect, 154
infidelity, 136–137
influencers, social media, 165, 167
inspiration, 213–214
Instagram, 164–166, 169
insulin, 40, 79
intensity, living with, 238–239
internet access, 183, 184
interpersonal therapy, 193
intimacy, 117, 132–133
Iolanthe (Gilbert and Sullivan), 115
Iowa Writers' Workshop, 220
iPhone, 166, 168n*, 191–192n*
isolation, 182
isomers, 47n*

J

Jami, Criss, 152
Jankowiak, William, 106
Jefferson, Thomas, 235n*
Jeong, Ken, 158
Jobs, Steve, 158
joy, from meaningful/purposeful living, 63, 229, 231, 238
justice, 236

K

kidney disease, 183n*
King, Ginevra, 229
Knight, Baker, 119n*
Knotts, Don, 213
K Street (Washington, D.C.), 177n*

L

Labash, Matt, 170
Lactobacillus acidophilus, 48
Lactobacillus casei, 48, 50
levodopa (L-dopa), 42, 46
Levoy, Gregg, 132, 143
Lexapro, 39
Lieberman, Daniel Z., xiii, 7, 19, 73, 217
life expectancy, 194
limbic system, 23, 24
limiting access
  to build anticipation for sex, 132–133
  to credit cards, 196–197
  to news, 175–179
  to online pornography, 188–189
  and separateness in relationships, 142–143
  to smartphones, 165–166
  to social media, 166–170
  treating dopamine-related problems by, 104
  to video games, 204
lion tamer mindset, 103–104
liraglutide, 79
lithium, 73
liver disease, 183n*
lobster, 55
Locke, John, 235n*
logotherapy, 91–93
loneliness, 110
Long, Michael E., xiii, 7, 217
long-term romantic relationships, 135–147
  anticipation of sex in, 132
  attachment in, 108
  beautiful lie in, 139–143
  being self-centered in, 137–139
  biologically-driven forces in, 123
  clearing bad feelings at end of, 146–147
  here-and-now in, 145
  marriage, 135–137
  new experiences in, 143–145
loot boxes, 200n*
love. *see* romantic love
love at first sight, 114–117
L-theanine, 46–47

INDEX

L-tyrosine, 46
Lubart, Todd, 213
lust, 107, 109–110, 136–137
lying
    about sexual behavior, 130
    the "beautiful lie," 139–143, 201

M

magic tricks, 209–210
Maher, Bill, 19
major depressive disorder, 31, 48
malaria, 77
mania, 31–32, 38, 44
marriage, 135–137
    components of love in, 136, 137
    dopamine subsystems and proposing, 25
    economic model of, 135–137
    legal aspect of, 135
    "me first" approach to, 137–139
    romantic love and, 106
    self-denial in, 137–138
    unhealthy attachment in, 111
masturbation, 130, 184, 186
mathematics, 158–159, 216–217
McCaffrey, Tony, 160
meaning, need for, 5–6, 91–92
meaningful or purposeful living, 227–239
    choosing to do good, 232–234
    dealing with pain, 231–234
    dissatisfaction with success and, 17
    and dopamine cycle, 225–226
    fulfillment from, 229–231
    hope and, 227–229
    joy from, 63, 229, 231, 238
    logotherapy and, 92–93
    pleasure and virtues in, 234–237
    reorienting toward, 237–239
median real income, 194
meditation, 110
meetings, in-person, 171
Meet the Press (TV series), 165
"me first" approach to marriage, 137–139
memory confabulation, 115
memory reconsolidation, 139–140
men
    closing time effect for, 118–121
    "dangerous"/risk-taking, 122–123
    problematic pornography use by, 182
    sexual attitudes of, 127–128
mental illness, 48, 215. *see also specific types, e.g.:* depression
mental management, 201

mental time travel, 24–26, 112
metabolite, 49n*
Michelangelo, 158
Middle Ages, 106
Midnight Cowboy (film), 213
millennial, 202
mindfulness
    in cue exposure therapy, 95
    and living in the present, 89–90
    to reduce problematic shopping, 193
    romantic love and, 110, 112
mind wandering, 210–211, 222
mobile devices. *see* cell phones
The Molecule of More (Lieberman & Long), xiii, 7, 217
mood, gaming to improve, 206
moral compromise, 130–131
motivation
    acceptance of dopaminergic, 85
    Adderall and, 43
    and desire dopamine, 23
    dopamine levels and, 37
    to exercise, 49–51, 69–71
    failure as, 63
    gut microbiome and, 49–51, 69–71
    need for meaning as, 91
    unhappiness and, 16
motor control, 22n*
MSK-DA01 implants, 76
Mulaney, John, 166, 218
Mull, Martin, 105
multiplayer games, 202
multitasking, 160–161, 222
music, 7–8, 215
"My Girlfriend, Who Lives in Canada" (song), 171

N

Nagarjuna (Akkineni Nagarjuna Rao), 158
naltrexone, 30
napping, 217–218
narratives, 93
Nasmyth, James, 158
National Football League (NFL), 183
National Institutes of Health, 46
natural environments, 214
natural selection, 152
NBC, 165
necessary dopaminergic activities, 60
neurons
    beta-carboline and, 73–74
    connections between, 211–213, 222

INDEX

transplanting, 75–77
NeuroSpark, 46
neurotransmitters. *see also specific types*
    for components of romantic love, 107
    defined, 19
    dopamine's interactions with other, 20–21
    electrical stimulation of receptors for, 72n*
    in love at first sight, 114
    placebo effect with ritual of treatment and, 98–99
    and romantic love, 106
    Xanax's effect on, 72
news consumption, 175–179
    author's experience with quitting, 175–178
    planning to quit, 178–179
    on social media, 169–170
Newton, Isaac, 211
NFL (National Football League), 183
Niemann, Christoph, 210
Nietzsche, Friedrich, 92
nihilism, 230, 237, 238
Nintendo Switch, 202
nootropic compounds, 40n*
norepinephrine, 107, 114, 181
normal dopamine levels, boosting, 39–51
    addition of dopamine to brain, 40–42
    manipulating the gut biome, 47–50
    nutritional supplements, 45–47
    overclocking the brain, 39–40
    understanding possibilities of, 50–51
    Wellbutrin or Adderall, 42–45
normalizing interventions, 39
Norman, Rodney Douglas, 62–63
Northcutt, Kent, 234
notifications, turning off, 166
Nottingham Trent University, 188–189
Nouwen, Henri, 233
novelty
    and creativity, 220–222
    in long-term romantic relationships, 143–145
    on social media, 168, 173
nucleus accumbens, 23, 30, 80
nutrition, 9–10, 50
nutritional supplements, 40, 45–47, 74–75

O

objective elements, 155
observing self
    to control PPU, 186
    in cue exposure therapy, 95
    to moderate gaming, 206
    for psychological flexibility, 88–89
    to reduce shopping behavior, 198
    in romantic love, 112
obsession, 195
obsessive-compulsive disorder, 83, 195
The Office (TV series), 19–20, 145
Ogden, Gina, 132
oniomania (compulsive shopping), 79, 193
online debates, 173–174
online pornography, 181–189
    disconnection of sexual pleasure from relationships with, 183–185
    magnitude of problematic use of, 181–182
    parallels between gaming and use of, 201
    tools to reduce use of, 185–189
open relationships, 136–137
opioids, 30
optimism, 19–20, 93, 139
oral contraceptives, 183–184
orgasm, difficulty achieving, 182, 189
originality, of ideas, 218
overdose, 29, 43
oxycodone, 79
oxytocin, 1, 107, 114, 182
Ozempic, 77

P

pain, dealing with, 231–234
panic, 83
Parker, Dorothy, 152, 210–211
Parkinson's disease, 22n*, 42
    beta-carbolines in, 73–74
    exercise motivation study of individuals with, 69–71
    neuron transplantation for treatment of, 76
Parks and Recreation (TV series), 175–176
passion, 132, 140–141, 143
Pasteur, Louis, 221
pattern recognition, 201, 204
peace, 33
penicillin, 73
Pennebaker, James W., 119, 120
Perel, Esther, 132
peripersonal space, 21, 138
Personalized Ads, on iPhones, 168n*
personal priorities, discussing, 117
persuasion, 174, 177–178
Pester, Patrick, 124–125

INDEX

PFC (prefrontal cortex), 72, 151
pharmaceuticals. *see also specific types and compounds*
    addressing addictive behaviors with, 193
    elimination of disease with, 77
    oral contraceptives, 183–184
    side effects of, 73–74
phase I trials, 76
phase II trials, 76
phase III trials, 76
phobias, 83
physical activity. *see* exercise and physical activity
physical space
    dedicated, for gaming, 205
    distancing yourself from trigger in, 188
    gamers in shared, 202
    sources of inspiration in, 213
pitch, 68
placebo effect, 40, 44–45, 47, 98–99
pleasure
    from achievement, 16–17
    dopamine and promise of, 236–237
    failing to find, 32
    in here-and-now (*see* here-and-now, finding pleasure in)
    inability to experience, 37
    in meaningful or purposeful living, 230–231, 234–237
    as motivator, 91
    virtues and appropriate, 235–237
politics, 175–177, 179
polyamory, 136–137
Pompa, Robin Nixon, 124–125
pornographic videos, 184
pornography, 131. *see also* online pornography; problematic pornography use (PPU)
pornography blockers, 185–186
positive illusion, 141
"possibility versus probability" test, 83
posttraumatic stress disorder (PTSD), 83
power, as motivator, 91
PPU. *see* problematic pornography use (PPU)
practice, creativity and, 211–213, 220
prebiotics, 50
precursor neurons, transplanting, 75–76
prefrontal cortex (PFC), 72, 151
preparation stage, in compulsive buying disorder, 195
present, living in, 89–90. *see also* here-and-now, finding pleasure in
prioritization, Eisenhower matrix for, 155–157

privacy, strangers', 171–172
probiotic supplements, 48–50
problematic pornography use (PPU)
    drug/alcohol addiction vs., 181, 182
    prevalence of, 182, 183
    regaining control over, 185–189
problem-solution approach to therapy, 86–87
problem solving
    task switching and, 161
    while falling asleep, 216–218
productivity, 149–161
    ambition and, 149–150
    building capacity for, 152–157
    divergent and convergent thinking for, 150–157
    multitasking and, 160–161
    skill in math and art and, 158–159
progressive muscle relaxation, 95
pros and cons lists, 155
prostitution, 131
PS5, 202
psychoanalysis, 97
psychological flexibility, 87–91
psychosis, 38, 44
psychotherapy. *see* therapy
psychotropic compounds, 40
PTSD (posttraumatic stress disorder), 83

Q
quantitative reasoning, 222
Queens University (Kingston, Ont.), 164

R
Rao, Akkineni Nagarjuna (Nagarjuna), 158
rapport, 117, 154
reactance, 119–121
reaction, in ACT, 88, 89
reality, hope and, 227–228
rebooting, 188–189. *see also* abstinence
recovery, 60
relationships. *see also* friendships; long-term romantic relationships
    disconnection of sexual pleasure from, 183–185
    effect of pornography on, 182
    face-to-face interactions in, 171
    with risk-taking men, 121–124
Remote Associates Test, 211–212
reproduction
    as biological imperative, 105, 108, 122
    competitive aspect in, 128
    romantic love and, 105, 108

INDEX

uncoupling of sex from, 183–184
resilience, building, 99
resistance
    to change, 60
    to premature closure, 219
    to sexual urges, 129–130
respiratory disease, 183n*
restlessness, 3–4
retail therapy, 191, 196–198
reverse reasoning, 154–155
reward prediction error, 26–28, 113–115, 125
Rice, Condoleezza, 158
the "right" person
    appeal of dangerous men, 121–124
    closing time effect, 118–121
    creating closeness with, 116–118
    first meeting with, 113–114
    and love at first sight, 114–116
    method for recognizing, 124–126
    searching for, 113–126, 200–201
risk-taking, 38, 121–124
ritual of treatment, 98–99
romantic love, 105–112. *see also* long-term romantic relationships
    components of, 107–109, 136, 137
    at first sight, 114–117
    H&N transmitters and being in love, 124
    in marriage, 136, 137
    memory reconsolidation in, 139–140
    normalcy of problems with, 233
    parallels between gaming and, 200–201
    role of biology in experience of, 105–107
    taming dopamine associated with, 109–112
    tests to identify, 124–125
romantic vacations, 143
Romeo and Juliet (Shakespeare), 142
runner's high, 57
Rybelsus, 77

S

sad feelings, accepting, 146
salience, 150, 153–154, 221
Saroyan, William, 238
satisfaction. *see also* dissatisfaction
    in meaningful/purposeful living, 233–234, 235
    overestimating, 33
    therapy to improve, 82, 83, 87
    and value identification, 90
Sayre, Zelda, 229
schedule, for gaming, 203–204
schizophrenia, 215
scientific consensus, 97–98
screentime, dopamine fasting and, 57
Screen Time app, 166
security, appeal of "dangerous" men and, 122
seizures, 75
selective serotonin reuptake inhibitors (SSRIs), 42
self-beliefs, 88
self-centered, being, 137–139
self-denial
    in dopamine fasting/revitalization, 60, 63
    and happiness vs. survival, 16
    in marriage, 137–138
    rehearsing moment of temptation and, 206
    to sustain sexual connection, 132–133
self-directed therapy, 82, 83
self-discipline. *see* discipline
self-disclosure, 117, 118
self-esteem, low, 194–195
self-reflection
    after a relationship ends, 146, 147
    on attachment, 111–112
    on attraction, 110–111
    on being in love, 125
    on gaming, 206–207
    on lust, 109
    on pornography use, 187–188
    on sexual behaviors, 129–131
    on shopping behavior, 196–197
    on virtues, 236–237
    on your admirable qualities, 147
self-responsibility, 138–139
semaglutide, 30n†, 77–80
sensation-based preferences, 106
Sepah, Cameron, 54
separateness, in long-term relationships, 142–143
separation, 111, 112, 132
serotonin, 1, 9, 42, 107, 114
sexual advertisement, 122–123
sexual attitudes, of men vs. women, 127–128
sexual behavior (sex), 127–134
    and competition, 127–128
    disconnection of, from relationships, 183–185
    dopamine levels and interest in, 37
    identifying problematic, 128–131
    increasing anticipation of, 132–133
    living in the present during, 89–90
    lust and, 108, 109

INDEX

natural limitations on, 43
normalcy of problems with, 233
other imperative-driven urges vs. urge for, 128, 133–134
outside of marriage, 136–137
parallels between gaming and, 200–201
with risk-taking/dangerous men, 121–123
uncoupling reproduction from, 183–184
sexual desire, pornography use and, 182, 189
sexual dysfunction, 182, 189
Shakespeare, William, 62, 142
Shaw, George Bernard, xv, 215
shock therapy. *see* electroconvulsive therapy (ECT)
shopping, 191–198
    benefits of consumerism, 193–194
    compulsive buying disorder, 194–196
    identifying problematic, 191–193
    online, 63
    overlapping solutions for problems with, 104
    techniques for reducing, 196–198
shopping stage, in compulsive buying disorder, 195
short-term relationships, with risk-taking men, 121–124
silence, 132
Simon, Carly, 142
sleep, 7–8, 154, 215
slot machines, 27, 168, 200
small changes, during dopamine fasting, 58
smartphone access, 165–166. *see also* cell phones
smart watch, ability to pay with, 196
smiling, 57
social interactions
    Adderall and, 43
    dopamine fasting and, 57
    face-to-face, 171, 172
    moderating gaming with, 201–202
    reward from, 117
    on social media, 170–174
social isolation, 182
social media, 163–175
    approval on, 178
    attention on, 63
    empty stimulation with, 168–170
    gaming vs. use of, 201
    hijacking of dopamine system by, 163–165
    interactions with strangers on, 170–174
    locking yourself out of, 166–167
    overlapping solutions for problems with, 104
    reducing feedback to algorithm for, 167–168
    reward prediction error with, 27
    technology to decrease use of, 165–166
social reward, 118
soft fascination, 214
soft stop, for gaming, 203
"Sorry, But Your Soul Just Died" (Wolfe), 106n*
special event, gaming as, 205
spending stage, in compulsive buying disorder, 195
spinal injection of dopamine, 41n†
Springsteen, Bruce, 136
SSRIs (selective serotonin reuptake inhibitors), 42
Stanford–Binet Intelligence Scales, 219
state of mind, for creativity, 220–222
"Stayin' Alive" (song), 67
Steam, 203
stillness, practicing, 8–9
Stills, Stephen, 120n*
stimulants, 43–45
stimulus, reaction and, 88
strangers, on social media, 170–174
stress reduction, 186, 206
stroke, 50, 183n*
subjective elements, 155, 218, 222
success, dissatisfaction and, 16–17. *see also* dopamine cycle [dopamine chase]
sugar, abstaining from added, 55, 60, 61
suicide, 31
Sullivan, Arthur, 115
Sullivan, Thomas, 158
surfing, in cue exposure therapy, 95
surprise, 8n*, 26–28, 144, 168
Survey Center for American Life, 182
survival
    as biological imperative, 106, 108
    as challenge, 63
    choosing between happiness and, 15–16
    evolution of brain to ensure, 18
    music and pathways associated with, 8
    threats to, 2, 3, 18
switch-up, during intimate sessions, 133

T

Talese, Gay, 136–137
task switching, 160–161, 222
Taylor, Shelley, 141

INDEX

tDCS (transcranial direct current stimulation), 72–73
technology
    and compulsive shopping, 193
    to decrease social media use, 165–166
    problem-solution approach in, 86–87
    to reduce online pornography use, 185–186
temperance, 236
The Tempest (Shakespeare), 62
temptation, planning to deal with, 185–186, 196, 205–206
testosterone, 107
Tetris effect, 216
L-theanine, 46–47
therapy, 81–100
    acceptance and commitment, 84–91, 98, 99, 110, 187, 193, 206
    addressing shopping behavior in, 192–193
    behavioral, 84, 86
    cognitive behavioral, 82–86, 99, 193, 197–198
    components of, 82
    to deal with root cause of PPU, 187
    exposure, 93–97
    logotherapy, 91–93
    selecting approach to, 97–99
    self-directed, 82, 83
    self-discipline and seeking, 99–100
theta waves, 67
third-party apps, for locking yourself out of social media, 167
Thy Neighbor's Wife (Talese), 136–137
time limits, for gaming, 203
Titanic, sinking of, 160
TMS (transcranial magnetic stimulation), 71, 72
tolerance
    dopamine revitalization and, 54, 61, 62
    to dopamine system stimulation, 28–31
    to stimulants, 43, 44
Torrance Test of Creative Thinking (TTCT), 218–219
tragic optimism, 93
transcranial direct current stimulation (tDCS), 72–73
transcranial magnetic stimulation (TMS), 71, 72
trauma, 182, 187
triggering events, for online pornography use, 187–188
trolling, social media, 165

TTCT (Torrance Test of Creative Thinking), 218–219
Tweedy, Jeff, 230
Twilight of the Idols (Nietzsche), 92
Twitter, 170
L-tyrosine, 46

U

unconscious mind, 210–211, 217, 221
University of Pennsylvania Perelman School of Medicine, 49
University of Virginia, 119
upregulation, 55, 56, 58
urgency, on Eisenhower matrix, 156–157

V

vacations, 143, 204
value identification, 90, 109
vasopressin, 182
ventral tegmental area, 23
Verizon, 191–192n*
Victoria's Secret, 142
video chats, 187
video game industry, 199
videos, pornographic, 184
violence, sex and, 136–137
virtues, 228, 232, 235–238
vitamin D, 41
vitamin K, 48
Vonnegut, Kurt, 174
vulnerability, 117, 118

W

wanting, having vs., 20–22
warfarin, 73
web browser, pornography blockers for, 185–186
Wegovy, 77
weight loss, 77–79
Weight Watchers, 78
well-being, positive illusion and, 141
Wellbutrin, 42–43
White, Chris, 234
Wilson, Rainn, 19–20, 26
Winnie-the-Pooh, 23
wisdom, 235–236
withdrawal, 30n*
Wolfe, Tom, 106n*
women
    attraction to risk-taking men for, 122–123
    closing time effect for, 118–121
    oral contraceptives for, 183–184

INDEX

　　problematic pornography use by, 182
　　sexual attitudes of, 127–128
work, dopamine fasting and, 56–57

X
X, 165
Xanax, 72, 164
Xbox, 202

Y
yoga nidra, 8–9

ABOUT THE AUTHOR

Trained as a physicist, Michael E. Long is co-author of the international bestseller *The Molecule of More*, which has been translated into more than twenty languages. As a playwright, he's had more than two dozen of his shows produced, most on New York stages. As a screenwriter, his honors include finalist for the grand prize in screenwriting at the Slamdance Film Festival. As a speechwriter, Mr. Long has written for members of Congress, US cabinet secretaries, presidential candidates, and Fortune 10 CEOs. A popular keynote speaker, Mr. Long has addressed audiences around the world, including at Oxford University. He teaches writing at Georgetown University, where he is a former director of writing. Mr. Long pursued undergraduate studies at Murray State University and graduate studies at Vanderbilt University.